LAW AND THE HUMAN BODY

Do you own your body? Advances in science and the development of genetic databases have added an aura of modern controversy to this long-standing and, as yet, unresolved problem. In particular, English law governing separated human tissue (including organs, DNA and cell-lines) is unsatisfactory. Despite the enactment of the Human Tissue Act 2004 UK, it remains uncertain what property rights living persons can claim over tissue separated from their bodies. The development of clear legal principles is necessary to protect the rights of individuals while also enabling the efficient use of such materials in medical research.

Part I of *Law and the Human Body* traces the evolution of English, Australian, United States and Canadian law in relation to human tissue separated from living persons and dead bodies. This includes a comprehensive examination of the Human Tissue Act 2004 UK as well as prominent judicial decisions, including *Re Organ Retention Group Litigation* [2005] QB 506, *Colavito v New York Organ Donor Network Inc* 8 NY 3d 43 (NY CA 2006) and *Washington University v Catalona* 490 F 3d 667 (8th Cir 2007). Analysis demonstrates that, although property rights and non-proprietary interests in separated human tissue are recognised in limited circumstances, no principled basis has been accepted either at common law or by statute for the recognition of these rights and interests. Part II of this book develops and defends a principled basis in English law for the creation and legal recognition of property rights and non-proprietary interests in separated human tissue. Significantly, the analysis and principles presented in *Law and the Human Body* have application across common law and civil law jurisdictions worldwide.

Law and the Human Body

Property Rights, Ownership and Control

Rohan Hardcastle

·HART·
PUBLISHING
OXFORD AND PORTLAND, OREGON
2007

Published in North America (US and Canada) by
Hart Publishing
c/o International Specialized Book Services
920 NE 58th Avenue, Suite 300
Portland, OR 97213-3786
USA
Tel: +1 503 287 3093 or toll-free: (1) 800 944 6190
Fax: +1 503 280 8832
E-mail: orders@isbs.com
Website: www.isbs.com

Hart Publishing, 16C Worcester Place, OX1 2JW
Telephone: +44 (0)1865 517530 Fax: +44 (0)1865 510710
E-mail: mail@hartpub.co.uk
Website: http://www.hartpub.co.uk

British Library Cataloguing in Publication Data
Data Available
ISBN: 978-1-84113-601-1 (hardback)

Typeset by Hope Services, Abingdon
Printed and bound in Great Britain by
TJ International Ltd, Padstow, Cornwall

To Coco

ACKNOWLEDGEMENTS

This book grew out of a doctoral thesis I completed at the University of Oxford with the generous assistance of the Menzies Memorial Foundation and Magdalen College. As with a generation of Oxford doctoral students, I was fortunate to receive guidance and inspiration from Professors Jim Harris and Peter Birks. Initially, Jim Harris was my supervisor. Subsequently, Peter Birks took on this role as the thesis evolved to finality. Jim and Peter were warm and generous individuals to whom I will always be indebted. Tragically, both died in 2004. I am particularly grateful to Professors Ewan McKendrick and Robert Chambers for supervising the final stages. Professor Margot Brazier and Roderick Bagshaw examined the thesis and have continued to provide valuable feedback.

I also collected a wonderful group of friends at Oxford who read numerous drafts and debated this project with me. These include Greg O'Mahoney, Danny Priel, Stephen Daley, Sue Gibbons and Ben Kremer. Thank you also to my college supervisors, Professor Colin Tapper and Katharine Grevling, for their advice throughout my time at Magdalen College.

After Oxford, I worked as an associate to Justice Dyson Heydon AC of the High Court of Australia. Justice Heydon's private law jurisprudence provides a fine example of the standard to which academic scholarship should aspire. I am extremely grateful for his guidance and support.

Others have provided generous assistance along the way. Thank you to Professor Lori Andrews, Jeremy de Beer, Prue Bindon, Marjorie Daley, James Edelman, Justice Robert French, Anne Gaustad, Professor Peter Handford, William Heath, Jonathon Herring, Jane Kaye, Professor Graeme Laurie, Bryan Leach, Kathleen Liddell, Kane Loxley, Roger Magnusson, Professor Paul Matthews, Jackie Milijash, Professor Stephen Munzer, Dianne Nicol, Julia Paolitto, Tina Piper, Eloise Scotford, Professor Loane Skene, Rebecca Skloot, Caroline Spruce, Professor Niall Whitty, Houda Younan and Jennifer Younan.

It has been a pleasure to work with Richard Hart and the staff at Hart Publishing—their professionalism has been evident throughout the publishing process.

Finally, nothing would have been possible without my family. I owe them more than thanks. In particular, my mother and father—Helen and Philip Hardcastle—were and continue to be a constant source of strength and inspiration.

Rohan Hardcastle
22 August 2007

TABLE OF CONTENTS

TABLE OF CASES

Australia

Canada

European Court of Human Rights

Germany

New Zealand

United Kingdom

United States of America

TABLE OF LEGISLATION, RULES, EUROPEAN INSTRUMENTS AND TREATIES

Australia

Canada

Denmark

European Union

Conventions

Directives

Germany

Iceland

International

United Kingdom

Statutes

Codes of Practice

Statutory Instruments

United States of America

1

Introduction

Do you own your body? Although the use of the human body as a medical resource is not a new development, advances in science and the development of genetic databases have made this question a contemporary controversy. Rapid scientific advancements are transforming the ability of scientific research to develop diagnostic and therapeutic products for complex diseases. Biological materials[1] are a central part of this process, which in turn has transformed such materials into valuable commodities. The law has not, however, developed clear principles as to what legal rights exist for individuals from whom biological materials are taken (sources) and those who remove and retain biological materials (recipients). In England, for example, it is not clear what property rights sources can claim over biological materials, despite the attempt at comprehensive regulation by way of the Human Tissue Act 2004 UK (HTA 2004). The development of clear legal principles is necessary so that individuals can have effective control over biological materials separated from their bodies and to enable the efficient use of such materials in medical research. The ambition of this book is to provide a foundation for the creation and allocation of legal rights (proprietary and non-proprietary) to human biological materials. Although the focus here is on English law, the book also draws on comparative material from other common law and civil law jurisdictions (such as the United States, Australia, Canada, Germany and Scotland). Significantly, the analysis and principles presented in this book have application across common law and civil law jurisdictions worldwide.

There are powerful and intersecting individual, medical and societal interests in this area. Individual sources have two broad interests in biological materials removed from their bodies. The first is an interest in the economic value of any profits that may be derived from separated biological materials (commercial participation). Profits have been and will continue to be made by researchers and companies that use biological materials to develop therapeutic products. The second is an interest in controlling the use and disposal of separated biological materials. There are at least four situations in which having a power of control may be significant. First, individuals may wish to determine the forms of scientific and medical research for which their biological materials are used. Empirical data

[1] The term 'biological materials' is defined in this book to mean any collection of human cells or sub-cellular components separated from a human body. This definition includes, for example, solid organs, blood components and deoxyribonucleic acid (DNA). Gametes and embryos are excluded from this definition of 'biological materials'.

indicates that the general public is concerned with controlling what research is performed on biological materials.[2]

Secondly, individuals may not wish separated biological materials to be used in a commercial setting.[3] The HTA 2004 authorises (in specified circumstances) the selling and buying of 'controlled material', which in broad terms is defined as human cells and body parts intended to be used for transplantation.[4] For example, 'controlled material' may be transferred from a DNA databank for the purposes of developing commercial products. Individuals may have moral, religious, philosophical or other objections to the commercial sale of separated biological material.

Thirdly, individuals may not wish separated biological materials to be immortalised.[5] In the normal course of scientific research, biological materials are analysed as part of a large collection. No particular sample is likely to receive close individual attention or special scrutiny. In the event that a cell-line is created the biological materials are likely to be the subject of a more focused and direct analysis. Individuals may not wish such immortalisation to occur for a number of reasons, including on moral or religious grounds.

Finally, control of biological materials may be significant for individuals when such materials are used to obtain personal genetic information. The protection of privacy is a matter of concern to the public, especially in the context of DNA databanks such as the UK Biobank.[6] Potential threats to individual privacy arise in several contexts. An individual's personal genetic information may contain information about which that individual does not wish to be made aware, such as, for example, predisposition to a genetic disease.[7] In the context of genetics, this right has been termed the 'right not to know'.[8] While this right is not an established legal principle in English law, disclosure of such information to the individual could constitute an invasion of that person's privacy. A further (and perhaps more substantial) privacy

[2] T Caulfield, 'Biobanks and Blanket Consent: The Proper Place of the Public Good and Public Perception Rationales' (2007) 18 *King's Law Journal* 209, 218–22; MR Robling, 'Public Attitudes Towards the Use of Primary Care Patient Record Data in Medical Research Without Consent: A Qualitative Study' (2004) 30 *Journal of Medical Ethics* 104, 105; ML Goodson and BG Vernon, 'A Study of Public Opinion on the Use of Tissue Samples from Living Subjects for Clinical Research' (2004) 57 *Journal of Clinical Pathology* 135, 137.

[3] D Green et al, 'Obtaining Informed Consent for Genetic Studies' (2006) 164 *American Journal of Epidemiology* 845, 849; C Williams, 'Australian Attitudes to DNA Sample Banks and Genetic Screening' (2005) 21 *Current Medical Research and Opinion* 1773, 1773.

[4] HTA 2004, s 32. For the statutory definition of 'controlled material' see ss 32(8), (9) and (10) and n 80 of ch 4 and accompanying text.

[5] As described below at nn 60–64 and accompanying text, immortalisation occurs through the creation of a cell-line.

[6] For a discussion of the UK Biobank, see nn 32–42 below and accompanying text. For a discussion of public concerns, see: K Korts, 'Genetic Databases and Public Attitudes: A Comparison of Iceland, Estonia and the UK' (2004) 8 *Journal of the Humanities and Social Sciences* 131, 141–42; Human Genetics Commission, *Public Attitudes to Human Genetic Information—People's Panel Quantitative Study Conducted for the Human Genetics Commission* (Department of Health, London, 2001) 5–6.

[7] Human Genetics Commission, *Inside Information* (Department of Health, London, 2002) [3.3].

[8] Australian Law Reform Commission, *Essentially Yours: The Protection of Human Genetic Information* (ALRC 96, Australian Government Publishing Service, Canberra, 2003) [7.24]. See, also: GT Laurie, *Genetic Privacy* (Cambridge, CUP, 2002) 264–66.

concern is the unauthorised disclosure of personal genetic information to third parties. An individual may wish to prevent the disclosure of genetic information to third parties, such as employers or insurers. These examples highlight potential infringements of individual privacy stemming from the use of biological materials.

The central question of this book concerns what legal principles should apply to protect these pecuniary and non-pecuniary interests. This involves first considering whether English law does and/or should recognise that sources have property rights with respect to separated biological materials. Some scholars argue that the law should recognise a source's property rights to their biological materials.[9] Others contend that a source's property rights are not justified at law but that the law should instead recognise a recipient's property rights.[10] The second question therefore becomes whether the English law of wrongs, including the law of tort, can protect an individual's interests in separated biological materials.[11]

This introductory chapter has three purposes. First, it outlines how biological materials are used for scientific and medical research in the context of biological material collections and genetic databases. It provides an introduction to the different physical forms of biological materials as they are used in scientific research. These different physical forms generate separate issues which have often been overlooked by legal academic commentary. It is important, therefore, to introduce the scientific principles and biological processes that explain these forms. Secondly, this chapter sets out the basic legal problem and introduces important legal concepts, including a definition of property rights. The chapter concludes by outlining the structure of the book.

A Biological Materials and Scientific Research

1 Human Tissue Collections and Genetic Databases

Biological materials are now valuable commodities. During the House of Commons Debate on 15 January 2004 (when the HTA 2004 had its Second Reading), Mr Frank Dobson observed:

[9] JK Mason and GT Laurie (eds), *Mason and McCall Smith's Law and Medical Ethics* (Oxford, OUP, 7th edn, 2005) 523 (England); C Hammond, 'Property Rights in Human Corpses and Human Tissue: The Position in Western Australia' (2002) 4 *University of Notre Dame Australia Law Review* 97, 113 (Australia); B Dickens, 'The Control of Living Body Materials' (1977) 27 *University of Toronto Law Journal* 142, 183 (Canada).

[10] JA Bovenberg, *Property Rights in Blood, Genes and Data* (Leiden, Martinus Nijhoff Publishers, 2005) 147; L Skene, 'Proprietary Rights in Human Bodies, Body Parts and Tissue' (2002) 22 *Legal Studies* 102, 123–27.

[11] The distinction between the law of wrongs and the law of tort within English jurisprudence is noteworthy. A wrong is defined as being a breach of a legal duty owed to a claimant: P Birks, 'Definition and Division: A Meditation on *Institutes 3.13*' in P Birks (ed), *The Classification of Obligations* (Oxford, OUP, 1995) 23. The law of tort falls within the law of wrongs. The analysis in this book is not limited to the law of tort but analyses the broader category of the law of wrongs.

> We cannot over-emphasise the importance of medical research on human tissues for the rest of us. For a start, such research helps us to identify specific illnesses and conditions. After such conditions are identified, the research enables professionals to understand their causes, and it is only after the causes are understood that professionals may come forward with a basis for the prevention or treatment of conditions. Medical research saves lives and relieves pain and suffering, and we must remember that it has gone on since time immemorial—from the Greeks studying anatomy to William Harvey discovering the circulation of the blood.[12]

In England, National Health Service (NHS) hospitals possess tens of thousands of organs and enormous collections of blocks and tissue slides.[13] Large collections also exist in Australia.[14] In the US, it is estimated that there are over 350 million human biological materials in storage.[15] Moreover, the scientific and commercial value of biological materials has grown significantly since the completion of the map of the human genome by the Human Genome Organisation and Celera in 2000.

(a) Tissue Collections

Throughout the world there are large numbers of human biological material collections. These collections were not necessarily established and maintained for research purposes. Instead, the collections typically have been established by: (1) pathology laboratories conducting medical testing; (2) health authorities conducting newborn screening procedures; and (3) tissue banks holding tissue and organs for transplantation.[16] Technological developments have made these collections a valuable resource for researchers. Often these biological materials were stored originally without any consent having been sought or given as to their future use. Consequently, the legal status and rights over them remain undefined. Even collections that have been set up more recently do not define the legal status of such materials. This includes, for example, the UK Stem Cell Bank, which provides a repository for stem cell-lines and facilitates the distribution of quality controlled stem cell-lines by the clinical and research communities.[17]

[12] *Hansard* HC vol 416 col 1003 (15 January 2004).

[13] M Brazier, *Medicine, Patients and the Law* (London, Penguin Books, 3rd edn, 2003) 475.

[14] G Clark, W Lipworth, L Bokey, JM Little and IH Kerridge, 'An Empirical Study of Tissue Banking in Australia: Navigating Regulatory and Ethical Challenges' (2006) 14 *Journal of Law and Medicine* 102, 102.

[15] JF Merz, 'On the Intersection of Privacy, Consent, Commerce and Genetics Research' in BM Knoppers (ed), *Populations and Genetics: Legal Socio-Ethical Perspectives* (New York, Kluwer Legal International, 2003) 257.

[16] For a detailed analysis of these contexts, see: *Essentially Yours*, above n 8, at [19.7]–[19.20].

[17] The UK Stem Cell Bank commenced in January 2003, and in September 2006, the bank announced that distribution stocks of the first four stem cell-lines (Shef-1, Shef-2, Shef-3 and hES-NCL-1) were available: http://www.ukstemcellbank.org.uk (accessed on 1 June 2007). The UK Stem Cell Bank issued a code of practice on 10 January 2007, which outlines the principles that will govern the operation of the bank: *Code of Practice for the Use of Human Stem Cell Lines* (available at http://www.ukstemcellbank.org.uk/code.html, accessed on 1 June 2007).

(b) Genetic Databases

Since the completion of the Human Genome Project in 2000, many specific population genetic databases have emerged. They contain collections of biological materials, information gathered from those materials and additional genetic information. These databases go by a variety of names: DNA databanks, gene banks, tissue banks, biobanks and genetic databases.[18] They may be public or private (or involve public–private collaborations), and often contain collections of both genetic sequence information and biological materials. 'Genetic databases' are defined here to include the wide variety of DNA databanks, gene banks, tissue banks and biobanks. Blood banks or databases established for forensic purposes are excluded from this definition.[19]

Two main forms of genetic databases have emerged in the last decade. The first group focuses on research for specific diseases and includes, for example, the English Acute Coronary DNA Library Project (ADLIB).[20] ADLIB aims to identify the genetic determinants of premature familial coronary artery disease. The genetic database contains a large collection of deoxyribonucleic acid (DNA) and white blood cell samples. In Australia, the Western Australian Research Tissue Network collects biological materials in relation to research for various forms of cancer.[21] In the US, PXE International (a disease advocacy group) has set up and maintained a privately held PXE Blood and Tissue Bank (PXE Bank), to assist research initiatives focused on the biology and genetics of pseudoxanthoma elasticum (PXE).[22] The PXE Bank contains biological materials and is utilised as a resource of materials for ongoing scientific research into PXE. PXE uses contractual arrangements to create property rights in biological materials that will be used by researchers.[23] Under the terms of its contracts, PXE retains any property rights to the biological materials.[24]

[18] House of Lords Select Committee on Science and Technology, *Human Genetic Databases: Challenges and Opportunities* (London, HMSO, 2001) Ch 3.3. For a discussion of these different terms, see: SMC Gibbons et al, 'Governing Genetic Databases: Challenges Facing Research Regulation and Practice' (2007) 34 *Journal of Law and Society* 163, 172–76.

[19] The primary purpose of forensic databases, such as the National DNA Database maintained for the police in England by the Forensic Science Services, is to create collections of genetic profiles for use in crime prevention and investigation. The physical samples, although having an important initial function, do not play a significant role once genetic information is extracted from them. Accordingly, forensic databases fall outside the scope of this book.

[20] http://www.dna-network.ac.uk/Studies/I00-I99/I20-I25/ (accessed on 1 June 2007).

[21] http://www.waimr.uwa.edu.au/etc/subpage.cfm/SID/11/PID/64/SPID/156 (accessed on 1 June 2007).

[22] SF Terry, 'Learning Genetics' (2003) 22 *Health Affairs* 166, 170; SF Terry and CD Boyd, 'Researching the Biology of PXE: Partnering in the Process' (2001) 106 *American Journal of Medical Genetics* 177, 178.

[23] DM Gitter, 'Ownership of Human Tissue: A Proposal for Federal Recognition of Human Research Participants' Property Rights in Their Biological Material' (2004) 61 *Washington and Lee Law Review* 257, 315.

[24] Terry and Boyd, above n 22, at 178.

The second type of database is intended more broadly to serve as a source of raw material for wide-ranging research.[25] Genetics research has traditionally focused on rare disorders caused by mutations in single genes. More recently, research has expanded to search for 'normal' gene variants—polymorphisms—that increase susceptibility to common diseases.[26] Identifying these polymorphisms requires large population studies.[27] Databases exist (or are being established) on a national level in England, the US, Estonia, Iceland, Singapore, Latvia, Italy, Sweden and Tonga. Scientists in the US are attempting to secure funding for a biobank project similar to the UK Biobank project. The US National Institute of Health (NIH) wants to collect genetic and medical information of 500,000 volunteers.[28] Without detailing the forms of research, these databases can be utilised for: (1) linkage studies; (2) association studies; (3) genetic epidemiology studies; and (4) pharmacogenetic studies. Two examples illustrate the emerging trend.

(i) Icelandic Health Sector Database

The Icelandic genetic database is one of the first collections to utilise biological materials on a national scale. Under the Act on a Health Sector Database (HSD),[29] the Icelandic government granted a licence to deCODE Genetics in January 2000 to construct a centralised database linking medical records with genealogical and genetic information.[30] The HSD regulates the creation and operation of a database of clinical medical information about Icelanders derived from their medical records. Separate legislation—the Icelandic Biobanks Act[31]—established guidelines for the use of biological materials in scientific research. The legislation does not, however, address the question of rights that individuals may have with respect to biological materials nor does it address the legal status of the materials in question.

(ii) UK Biobank

The UK Biobank project is a joint initiative between the Medical Research Council, the Wellcome Trust, the Department of Health and the Scottish

[25] RJ Nederhand, 'Logistics and Quality Control for DNA Sampling in Large Multicentre Studies' (2003) 1 *Journal of Thrombosis and Haemostasis* 987, 987.

[26] The Welcome Trust Case Control Consortium, 'Genome-wide Association Study of 140,000 Cases of Seven Common Diseases and 3,000 Shared Controls' (2007) 447 *Nature* 661, 661; LR Cardon and JI Bell, 'Association Study Designs for Complex Diseases' (2001) 2 *Nature Reviews* 91, 91.

[27] Cardon and Bell, above n 26, at 91.

[28] http://www.sciencemag.org/content/vol304/issue5676/s-scope.shtml (accessed on 1 June 2007).

[29] Act on a Health Sector Database, No. 139/1998. For an English translation of this legislation, see: http://www.mannvernd.is/english/laws/law.HSD.html (accessed on 1 June 2007).

[30] HT Greely, 'Iceland's Plan for Genomics Research: Facts and Implications' (2000) 40 *Jurimetrics* 153, 171–72; MJ Smith, 'Population-Based Genetic Studies: Informed Consent and Confidentiality' (2001) 18 *Santa Clara Computer and High Technology Law Journal* 57, 68.

[31] Act on Biobanks, No.110/2000. For an English translation of this legislation, see: http://www.mannvernd.is/english/laws/Act.Biobanks.html (accessed on 1 June 2007).

Executive.[32] The project differs from the Icelandic HSD in that only a subset of the population will be sampled,[33] through a collection of DNA samples from 500,000 volunteers.[34] The UK Biobank intends to follow the health of volunteers for a number of years, collecting and linking information on environmental and lifestyle factors to medical records and biological materials.[35] The biological materials will be stored for future genetic analysis. Importantly, the project also proposes that some original biological materials be immortalised.[36] Immortalisation of human cells can be achieved through the creation of cell-lines.[37] The UK Biobank also intends that the resources (including biological materials and cell-lines) will be utilised by commercial and scientific users.

At present, there is no specific legislation governing the UK Biobank.[38] In September 2003, the UK Biobank Interim Advisory Group issued a 'Draft UK Biobank Ethics and Governance Framework', which set out the intended operating principles for the UK Biobank and the second version, entitled 'UK Biobank Ethics and Governance Framework' (Biobank Framework), was released in July 2006.[39] It is not clear what legal interests (if any) the Biobank Framework accords individuals to separated biological materials. Its provisions stipulate that UK Biobank will be the legal owner of the data and sample collections and, further, that participants will not retain any property rights in the materials.[40] Under the Biobank Framework, the UK Biobank is given the right to take legal action against unauthorised use or abuse of the database or materials and also retains the right to sell or destroy the materials.[41] Individuals will have a right to withdraw from participation, and their biological materials may then be destroyed.[42]

[32] See: http://www.ukbiobank.ac.uk (accessed on 1 June 2007).

[33] For a further discussion of the distinctions between the Icelandic Database and UK Biobank, see: J Targiel, 'Commercialisation of Human Bodily Substances and Human Genetic Information' (2006) 3 *Journal of International Biotechnology Law* 244, 256–62.

[34] See: http://www.ukbiobank.ac.uk (accessed on 1 June 2007).

[35] FC Rawle, 'UK DNA Sample Collections for Research' in BM Knoppers (ed), *Populations and Genetics: Legal and Social–Ethical Perspectives* (New York, Kluwer Legal International, 2003) 4–6.

[36] UK Biobank, *Sample Handling and Storage Protocol and Recommendations* (2004) (available at http://www.ukbiobank.ac.uk/docs/Sample_Handling_and_Storage_Subgroup_report_for_comment.pdf, accessed on 1 June 2007) 7, 19; Department of Health, *Protocol for the UK Biobank* (Department of Health, London, 2002) 19.

[37] Cell-lines are a specific (and significant) form of biological material and are discussed below at nn 60–66 and accompanying text.

[38] SMC Gibbons, 'Are UK Genetic Databases Governed Adequately? A Comparative Legal Analysis' (2007) 27 *Legal Studies* 312, 318–19.

[39] Available at http://www.ukbiobank.ac.uk/ethics/efg.php (accessed on 1 June 2007).

[40] Department of Health, *UK Biobank Ethics and Governance Framework* (Department of Health, London, 2006) 14.

[41] Ibid, at 14.

[42] Ibid, at 10–11. It is still unclear whether UK Biobank undertakes to destroy biological materials on participant withdrawal.

2 Biological Materials

Currently, the most widely used biological materials are derived from blood samples.[43] Such materials are utilised for various research purposes. Continuing advances in knowledge about human cells and improved means of analysis will ensure that biological materials continue to be a significant raw material in scientific research.[44]

(a) Physical Manifestations

It is important to understand the different physical forms of biological materials. To date, academic commentary has focused primarily on biological materials as they exist in their original form.[45] Insufficient attention has been paid to additional biological materials created during the course of medical or scientific research. Such materials have become essential tools in the development of medical therapies and diagnostics.[46] Accordingly, this book considers the legal status and principles applicable to these additional materials. As a preliminary matter, it is useful to explain the four fundamental forms of biological materials used in scientific processes.

(i) Original Biological Materials

Biological materials may exist in their original state, fixed and mounted. These materials are analysed to obtain information relating to their identity and behaviour. Although biological materials often are placed onto microscope slides, there is no standardised practice for storing them.[47] Such materials are commonly fixed in solution and mounted in paraffin blocks.[48] Before these materials are placed in the paraffin block, they are 'fixed'. Fixing involves immersing them in preservative so that they will harden for processing.[49] Water and fat are then removed and replaced with paraffin wax.[50] On average, a paraffin block will consist of approxi-

[43] RJ Nederhand, 'Logistics and Quality Control for DNA Sampling in Large Multicentre Studies' (2003) 1 *Journal of Thrombosis and Haemostasis* 987, 987. For example, the UK Biobank intends to use blood samples of 50 ml from 500,000 individuals: Department of Health, *UK Biobank Ethics and Governance Framework* 3.

[44] SJ Younger, MW Anderson and R Schapiro (eds), *Transplanting Human Tissue* (New York, OUP, 2004) 33; House of Lords Select Committee on Science and Technology (Fourth Report 1999) [3.4].

[45] See, for example: Skene, above n 10; G Dworkin and I Kennedy, 'Human Tissue: Rights in the Body and Its Parts' [1993] 1 *Medical Law Review* 291.

[46] JP Walsh, WM Cohen and A Arora, 'Licensing Research Tools and Biomedical Innovation' in S Merrill, R Levin and M Meyers (eds), *Innovation in a Knowledge Based Economy* (Washington, National Academy Press, 2003) 289.

[47] U Lehmann and H Kreipe, 'Real-Time PCR Analysis of DNA and RNA Extracted from Formalin-Fixed and Paraffin-Embedded Biopsies' (2001) 25 *Methods* 409, 409.

[48] Paraffin blocks are blocks of hard paraffin with biological materials embedded in them: Lehmann and Kreipe, above n 47, at 409.

[49] Retained Organs Commission, *Tissue Blocks and Slides: A Consultation Paper* (NHS, London, 2002) 4.

[50] Ibid.

mately 50 per cent biological material.[51] Microscope slides are then made by cutting shavings of biological materials from the paraffin blocks. The shaved material is then transferred to a glass slide, stained and secured by a self-hardening resin. The process of staining does not alter the physical composition of the biological materials, but instead makes the DNA structure in the materials more easily identifiable.[52]

(ii) Amplified DNA

Biological materials may take the form of DNA extracted from biological materials. DNA is a physical object. Before isolating the genomic DNA (gDNA), the cells must be lysed and the DNA made available for extraction.[53] The most common source of extracted DNA is blood.[54] Once the DNA has been isolated, it may also be amplified. Increasingly, scientists are using polymerase chain reaction (PCR) technology[55] to amplify a fragment of DNA from the original biological material (amplified DNA). For example, PCR has been successfully utilised in studying the epidemiology of Lyme disease.[56] More recently, scientists have reported the success of Multiple Displacement Amplification (MDA) in amplifying the entire DNA of a cell directly from biological materials.[57] MDA is a significant breakthrough in scientific research; previously, PCR technology had been limited to amplification of DNA fragments.[58] The MDA technique enables large amounts of DNA to be created from small amounts of biological material. The technology also has the potential to be applied to paraffin-embedded biological materials. If successful, this development would have significant implications for biological material collections throughout the world.[59]

[51] Independent Review Group, *Retention of Organs at Post-Mortem Final Report* (Edinburgh, 2001) (available at http://www.scotland.gov.uk/Publications/2003/11/18558/29413, accessed on 1 June 2007) 48.

[52] JW Tung, 'New Approaches to Fluorescence Compensation and Visualization of FACS Data' (2004) 110 *Clinical Immunology* 277, 278; A Orfao and A Ruiz-Arguelles, 'General Concepts About Cell Sorting Techniques' (1996) 29 *Clinical Biochemistry* 5, 8.

[53] HJ Schunemann, 'Short term Storage of Blood Samples and DNA Isolation in Serum Separator Tubes for Application in Epidemiological Studies and Clinical Research' (2000) 10 *Ann Epidemiol* 538, 538–39.

[54] Although DNA can be collected from buccal (cheek) swabs, studies have found that these cells are of limited utility because of the significant variation in yield and quality of the DNA obtained: LC Hartley, 'Collection of Buccal Cell DNA Using Treated Cards' (2000) 9 *Cancer Epidemiol* 501, 502.

[55] P Rabinow, *Making PCR: A Story of Biotechnology* (Chicago, University of Chicago Press, 1996). For a discussion of the use of this technology in forensic circumstances see: KT MacKnight, 'The Polymerase Chain Reaction (PCR): The Second Generation of DNA Analysis Methods Takes the Stand' (2003) 20 *Santa Clara Computer and High Technology Law Journal* 95, 134–48.

[56] DH Persing, 'Detection of Borrelia Burgdorferi DNA in Museum Specimens of Ixodes Dammini Ticks' (1990) 249 *Science* 1420, 1421.

[57] S Hosono, 'Unbiased Whole-Genome Amplification Directly from Clinical Samples' (2003) 13 *Genome Research* 954, 954; RS Lasken and M Egholm, 'Whole Genome Amplification: Abundant Samples of DNA from Precious Samples or Clinical Specimens' (2003) 21 *Trends in Biotechnology* 531, 532.

[58] Previously, there was a large variation in the extent to which this technique could produce a whole DNA amplification. Techniques suffered from amplification bias, meaning that regions of the genome were lost in the process: Hosono, ibid at 955.

[59] Lasken and Egholm, above n 57, at 534.

(iii) Cell-lines

Cell-lines are another form of biological material created from explants of the original biological materials. Each cell-line is derived from a progenitor cell, and is a self-replicating material grown outside the original material.[60] The HeLa cell-line is perhaps the most well-known cell-line and the original cells that were used to create the cell-line were taken from Henrietta Lacks.[61] Stem cell-lines are a more recent example, and they offer great potential scientific and therapeutic value for the future.[62] The importance of a cell-line lies in the uniformity of its biological material, allowing scientists to carry out repeatable experiments.[63] Cell-lines are consequently used in the study of biological processes and play an increasingly central role in testing the potentially harmful effects of drugs and other compounds on living matter.[64] In particular, studies of leukaemia-lymphoma cell-lines have provided useful insights into studies of hematopoietic malignancies.[65] Blood and bone marrow materials are the most commonly used biological materials to initiate the production of cell-lines. Other liquid specimens (such as pleural effusion, ascites or cerebro-spinal fluid) and solid biological materials (lymph node, tonsil or spleen) are employed less often.[66] Despite these developments, the success rate of creating a cell-line from biological materials is relatively low. As a result, cell-lines are valuable products and there is ongoing demand for novel cell-lines.

(iv) Modified Biological Materials

A fourth form of biological material are those altered in physical structure by the introduction of additional biological materials, such as the insertion of DNA from a different source. In scientific terminology this technique is referred to as the insertion of 'foreign DNA'. Foreign DNA is usually inserted into a cell-line to

[60] 'UKCCCR Guidelines for the Use of Cell Lines in Cancer Research' (2000) 82 *British Journal of Cancer* 1495, 1496–98.

[61] For the original scientific account of the HeLa cell-line, see: GO Gey, WD Coffman and MT Kubicek, 'Tissue culture studies of the proliferative capacity of cervical carcinoma and normal epithelium' (1952) 12 *Cancer Research* 264 and WF Scherer, JT Syverton and GO Gey, 'Studies on the propagation in vitro of poliomyelitis viruses. IV. Viral multiplication in a stable strain of human malignant epithelial cells (strain HeLa) derived from an epidermoid carcinoma of the cervix' (1953) 97 *The Journal of Experimental Medicine* 695, 695–710; See, also: R Skloot *The Immortal Life of Henrietta Lacks* (Crown, New York, forthcoming) ch 1.

[62] For a discussion of such potential, see: JB Spillane and MA Henderson, 'Cancer Stem Cells: A Review' (2007) 77 *ANZ Journal of Surgery* 464, 464; PA Beachy, SS Karhadkar and DM Berman, 'Tissue Repair and Stem Cell Renewal in Carcinogenesis' (2004) 432 *Nature* 324; R Lovell-Badge, 'The Future for Stem Cell Research' (2001) 414 *Nature* 88, 88–91.

[63] HG Drexler, Y Matsuo and RAF MacLeod, 'Continuous Hematopoietic Cell Lines as Model Systems for Leukemia-Lymphoma Research' (2000) 24 *Leukemia Research* 881, 885.

[64] Department of Health, *Human Bodies, Human Choices: The Law on Human Organs and Tissue in England and Wales* (Department of Health, London, 2002) [17.3].

[65] For example, it is estimated that there are over 1,000 leukaemia-lymphoma cell-lines: HG Drexler and RAF MacLeod, 'Leukaemia-lymphoma Cell Lines as Model Systems for Hematopoietic' (2003) 35 *Annuals of Medicine* 404, 405.

[66] Drexler, Matsuo and MacLeod, above n 63, at 889.

create a modified cell-line consisting of the original cell-line and the foreign DNA.[67] Foreign DNA also can be inserted into original (often termed primary) biological material. Commonly, viruses are used to transplant the foreign DNA inside the cells. Retroviruses—especially the subset of these called lentiviruses—are the most effective. The earliest example of the use of lentiviruses to introduce foreign DNA into primary human cells was reported in 1996.[68] While the use of retroviruses is not yet advanced in the context of human biological materials, a technique has been developed by which genetically-modified mouse cells can be re-injected safely into mice.[69] Cells are removed from the mouse and then inserted with foreign DNA so as to prevent the cells from causing an immune disease. Prior to the insertion of DNA, these cells have the potential to be aggressive and to attack the immune system. Importantly, when the cells (which have been subjected to the insertion of foreign DNA) are reinserted into the mouse, the function of the cells change. Specifically, the mouse cells lose the ability to cause immune system disease in the new host. Instead, the cells protect the mouse from immune system disease. The success of this technique in the context of mouse cells is significant because the next step is to transfer its application to human biological materials. Although the technique is not currently utilised in the context of human biological materials, rapid advances in immunology over the last decade are moving in this direction.

For the purposes of this book, where foreign DNA is inserted into an original cell or a cell-line the resulting biological materials will be referred to as 'modified biological materials'. Typically, the creation of these modified biological materials will also be subject to intellectual property rights, as they may constitute 'inventions'. These materials raise difficult legal issues. For example, at what point should the law recognise that biological materials have been transformed into a different physical object? If biological materials have been transformed into new objects then different property principles will apply. This is one of the key areas addressed in Part II. By way of definition, this book refers to amplified DNA, cell-lines and modified biological materials collectively as 'additional biological materials'.

(b) Genetic Information

This book is concerned primarily with biological materials as physical objects and not 'genetic information' extracted from biological material. This is because English law treats biological materials very differently from genetic information derived from such materials. The regulation of medical and genetic information

[67] For one of the pioneering works in this field see: KM Huttner, GA Scangos and FH Ruddle, 'DNA-Mediated Gene Transfer of a Circular Plasmid into Murine Cells' (1979) 76 *National Academy of Sciences USA* 5820, 5821.

[68] L Naldini, 'In Vivo Gene Delivery and Stable Transduction of Non-dividing Cells by a Lentiviral Vector' (1996) 272 *Science* 263, 264.

[69] S Hori, T Nomura and S Sakaguchi, 'Control of Regulatory T cell Development by the Transcription Factor Foxp3' (2003) 299 *Science* 1057, 1059.

and the question of whether such information can be the subject of property rights raise separate questions beyond the scope of this work.[70] Nonetheless, questions relating to genetic information are considered in addressing the principal question of the book and a definition of 'genetic information' is therefore necessary. There are many different forms of genetic information which make a single definition of the concept difficult. Information about an individual's skin colour or gender, for example, is arguably genetic information. More specific forms of genetic information include the results of DNA analysis. The term 'genetic information' is used here to mean information about an individual's genetic characteristics, as derived from their DNA, ribonucleic acid (RNA), genes and chromosomes.

B Legal Framework

The legal status of biological materials, and foundational legal concepts, such as property rights and bodily integrity, remain to be considered.

1 Legal Status of Biological Materials

Despite the increasing utilisation of biological materials in scientific research, it is uncertain whether leading common law jurisdictions recognise property rights in biological materials. In England, for example, neither existing legislation nor the common law currently recognises a source's property rights to separated biological materials. In general, English law has been guided by the 'no property' principle (the principle which declares that there is no property in corpses).[71] However, a small collection of court decisions have held that substances of the human body (such as blood and urine samples) can be classified as 'property' capable of being

[70] For example, it is not yet settled whether English law regards information as capable of a proprietary characterisation. The more accepted view is that information is not property: *Boardman v Phipps* [1967] 2 AC 46 (HL) 127–28 (Lord Upjohn). Instead, the right to control the use of personal information is governed by the doctrine of confidence and the Data Protection Act 1998. In *Douglas v Hello! Ltd (No 3)* [2005] EWCA 595; [2006] QB 125, the Court of Appeal stated at [119]: '[w]e have concluded that confidential or private information, which is capable of commercial exploitation but which is protected by the law of confidence, does not fall to be treated as property that can be owned and transferred'. Similarly, in *OBG Limited v Allan; Douglas v Hello! Ltd; Mainstream Properties Limited v Young* [2007] UKHL 21; [2007] 2 WLR 920, Lord Walker observed at [275] that 'information, even if it is confidential, cannot properly be regarded as a form of property'. See, also: *Farah Constructions Pty Ltd & Ors v Say-Dee Pty Ltd* [2007] HCA 22 (HCA) [118] (Gleeson CJ, Gummow, Callinan, Heydon and Crennan JJ). For a more detailed discussion of the classification of information as property, see: N Palmer, 'Information as Property' in N Palmer and E McKendrick (eds), *Interests in Goods* (London, LLP, 2nd edn, 1998) 3.

[71] *Exelby v Handyside (Handyside's Case)* (1749) 2 East PC 652, 653; *R v Kelly* [1999] QB 621, 630–31; RS Magnusson, 'Proprietary Rights in Human Tissue' in Palmer and McKendrick (eds), ibid, at 27–29.

stolen,[72] and thus suggest that property rights may exist in biological materials. Nevertheless, English law has not determined whether a living person can claim property rights to separated biological materials. Accordingly, legal commentators rightly point to a conceptual gap in the existing law.[73] Grubb notes that there is no clarity on the central questions: '[f]irst, does excised tissue fall within the property regime at all? Second, if it does, who is entitled to claim "property interests" in it? Is it the source or the remover?'[74] Part I of this book provides a detailed analysis of how the common law and statutory regimes have addressed questions concerning the human body and separated bodily materials.

2 Proprietary Protection

Before commencing this analysis, it is necessary to outline the features of property rights. Despite the ubiquity of property concepts in law, the nature and content of property rights are fraught with ambiguity. This ambiguity arises over a consideration of two issues: (1) the nature of property rights; and (2) objects over which property rights extend. No single theory has produced a satisfactory universal or general meaning as to the nature of a property right. Although there are numerous contemporary attempts to define private property rights,[75] it is not necessary to address this jurisprudential debate here. This book assumes a working definition of property and takes as its starting point the definition of 'property rights' as a bundle of different rights to a thing. 'Property' should be understood as referring to things that are the subject of property rights. This definition has been referred to by both the US Supreme Court[76] and the Australian High Court.[77] The US Supreme Court recently agreed with an earlier statement in *United States v General Motors*[78] that 'property . . . consists of "the group of rights which the so-called owner exercises in his dominion of the physical thing," such "as the right to possess, use and dispose of it"'.[79] In *Yanner v Eaton*, the Australian High Court

[72] *R v Rothery* [1976] RTR 550 (CA) 552–53 (blood specimen); *R v Welsh* [1974] RTR 478 (CA) 479 (urine sample); *R v Herbert* [1961] JPLGR 12, 13 (hair sample).

[73] D Morgan, *Issues in Medical Law and Ethics* (London, Cavendish, 2000) 83; JK Mason, RA McCall-Smith and GT Laurie, *Law and Medical Ethics* (London, Butterworths, 1999) 486; A Grubb, 'I, Me, Mine: Bodies, Parts and Property' (1998) 3 *Medical Law International* 299, 305; Magnusson, above n 71, at 28.

[74] Grubb, ibid, at 305.

[75] See, especially: JE Penner, *The Idea of Property in Law* (Oxford, OUP, 1997) 1–6; R Posner, *Economic Analysis of Law* (New York, Aspen, 5th edn, 1998) 39; JW Harris, *Property and Justice* (Oxford, Clarendon Press, 1996) 6–8; S Munzer, *A Theory of Property* (Cambridge, CUP, 1990) 15–36; J Waldron, *The Right to Private Property* (Oxford, OUP, 1988) 26; LC Becker, *Property Rights: Philosophical Foundations* (London, Routledge, 1977) 2–4.

[76] *United States v General Motors* 323 US 373; 65 S Ct 357 (1945); American Law Institute, *A Concise Restatement of Property Law* (Minn, American Law Institute Publishers, 2001) 1.

[77] *Yanner v Eaton* [1999] HCA 53; (1999) 201 CLR 351 (HCA) [17] (Gleeson CJ, Gaudron, Kirby and Hayne JJ).

[78] *General Motors*, above n 76.

[79] *Phillips v Washington Legal Foundation* 524 US 156, 170; 118 S Ct 1925, 1933 (1998).

observed that 'property does not refer to a thing; it is a description of a legal relationship with a thing. It refers to a degree of power that is recognised in law as power permissibly exercised over the thing'.[80]

Adopting this definition raises the question of the content of the bundle of rights. In his essay entitled 'Ownership', Honoré attempted to identify the rights that constitute the bundle, identifying eleven 'standard incidents' of a 'full liberal concept of ownership'.[81] The terms 'ownership' and 'owner' are often used in the context of property rights and they indicate the content of property rights. Ownership is, however, a complex term. Its complexity stems from the fact that ownership gives rise to a collection of rights which differ considerably in their character and effect from one situation to another.[82] Accordingly, jurists have found it difficult to define ownership precisely. The term ownership is used in this book to describe 'the greatest possible [bundle of rights] in a thing which a mature system of law recognizes', comprising a number of rights, incidents and liabilities.[83] This includes, for example, the right to sell the property.

The second question is, over what objects do property rights extend? The inability to find a universal definition of a property right in part stems from the question as to what objects (physical or otherwise) can be the subject of property rights. This book is concerned with property rights over physical biological materials and not intangible genetic information. As a result, the question of what objects property rights extend over assumes less significance, but nonetheless requires explanation. Two schools of thought purport to define the types of objects over which property rights may extend. The first is based on a narrow conception of property rights as rights in relation to a thing. The second is a broader conception that equates property rights with wealth.[84] This book adopts the first view.

The narrower conception of property rights focuses on the existence of a thing. Within this framework, the difficult question becomes: what is a 'thing'? Does it include both tangible and intangible objects? Or is a thing limited to an object that has a physical existence? Some commentators emphasise that (under the narrower conception) property rights require the existence of a physical entity.[85] Others argue that a physical thing is not necessary and that the concept can include

[80] *Yanner*, above n 77, at [17] (Gleeson CJ, Gaudron, Kirby and Hayne JJ).

[81] They are: (1) the right to possess; (2) the right to use; (3) the right to manage; (4) the right to income; (5) the right to capital; (6) the right to security; (7) the incident of transmissibility; (8) the incident of absence of term; (9) the duty to prevent harm; (10) liability to execution; and (11) residuary character. For an explanation of each right see: AM Honoré, 'Ownership' in AG Guest (ed), *Oxford Essays in Jurisprudence* (Oxford, OUP, 1961) 112–28. See, also: AM Honoré, 'Property and Ownership: Marginal Comments' in T Endicott, J Getzler and E Peel (eds), *Properties of Law: Essays in Honour of Jim Harris* (Oxford, OUP, 2006) 131–35.

[82] BA Rudden, 'Things as Thing and Things as Wealth' in JW Harris (ed), *Property Problems: From Genes to Pension Funds* (London, Kluwer Law International, 1997) 150.

[83] Honoré, 'Ownership' in Guest (ed), above n 81, at 162.

[84] Harris, above n 75, at 140. 'Wealth' is used in this book to refer to something which can be accorded a monetary value, such as a house, money and shares: Rudden, above n 82, at 87.

[85] B Boukhart, 'What is Property?' (1990) 13 *Harvard Journal of Law and Public Policy* 775, 796.

patents,[86] shares[87] and securities.[88] Although this debate raises difficult questions, it is clear that for a thing to be the subject of property rights it must be distanced from a human subject.[89] For example, the use of an individual's talents is not separable from the individual and therefore not capable of classification as a thing that can be subject to property rights. Similarly, it is not possible to say that we 'own our entire bodies'.[90] In *R v Bentham*,[91] the House of Lords considered whether a person who was using his hand to give the impression he was holding a firearm inside a jacket was charged with being in possession of an imitation firearm contrary to subsection 17(2) of the Firearms Act 1968 UK. The trial judge held that a person could be convicted on those facts and the Court of Appeal (Kennedy LJ, Curtis and Forbes JJ) upheld that decision.[92] The House of Lords reversed the Court of Appeal decision and quashed the conviction. Lord Bingham (with whom Lord Steyn, Lord Phillips, Lord Rodger and Lord Carswell agreed)[93] found that an unsevered hand or finger was not capable of being possessed, with the consequence that the charge under section 17(2) could not be sustained.[94] His Lordship reasoned:

> One cannot possess something which is not separate and distinct from oneself. An unsevered hand or finger is part of oneself. Therefore, one cannot possess it. Resort to metaphor is impermissible because metaphor is a literary device which draftsmen of criminal statutes do not employ. What is possessed must under the definition be a thing. A person's hand or fingers are not a thing.[95]

This statement confirms that biological materials which remain part of the human body are not things because they lack the requirement of separability. It also creates the potential for biological materials detached from a human body to be classified as things.

3 Non-proprietary Protection

The conceptual lacuna in the law concerning separated biological materials can be contrasted with the protection afforded by the law to invasions of an individual's

86 Honoré, 'Property and Ownership: Marginal Comments' in Endicott, Getzler and Peel (eds), above n 81, at 131.

87 FH Lawson and B Rudden, *The Law of Property* (Oxford, OUP, 3rd edn, 2002) 33–36; R Pennington, 'Can Shares in Companies Be Defined' (1989) 10 *Company Lawyer* 140, 144.

88 J Benjamin, *Interests in Securities: A Proprietary Analysis of the International Securities Markets* (Oxford, OUP, 2000) [14.31].

89 *R v Bentham* [2005] UKHL 18; [2005] 1 WLR 1057, [8] (Lord Bingham). See, also: Penner, above n 75, at 111; Harris, above n 75, at 332.

90 Penner, above n 75, at 125.

91 [2005] UKHL 18; [2005] 1 WLR 1057.

92 *R v Bentham* [2003] EWCA Crim 3751; [2004] 1 Cr App R 487.

93 *Bentham*, above n 91, at [12] (Lord Steyn), [13] (Lord Phillips), [14] (Lord Rodger), [15] (Lord Carswell).

94 Ibid, at [11] (Lord Bingham).

95 Ibid, at [8] (Lord Bingham).

own body. For centuries, the common law has recognised non-proprietary rights protecting invasions of the human body. Blackstone argues that there exists an absolute right to personal security vested in each person:

> [The] right of personal security consists in a person's legal and uninterrupted enjoyment of his life, his limbs, his body, his health, and his reputation.[96]

English law recognises that each individual has an innate right to bodily integrity, encompassing the right to exclusive possession and use of one's own body as against everyone else. The following analysis examines, first, how the right to bodily integrity is grounded in notions of autonomy and human dignity and, secondly, how the right to bodily integrity is protected by the law of tort and the criminal law.[97] The right to bodily integrity derives from the concept of autonomy,[98] which is the capacity that individuals have to act independently. The word 'autonomy' is derived from the ancient Greek for 'self' and 'law' and means literally 'the having or making of one's own laws'.[99] The concept of autonomy is also linked to the concept of human dignity.[100] Indeed, Lord Millett recently observed that autonomy is an important component of human dignity.[101]

The relationship between autonomy and human dignity has been the subject of extensive philosophical debate. In Roman times, the original meaning of dignity (*dignitas*) referred to an acquired social and political status, implying important personal achievements and moral integrity.[102] It thus represented a manifestation of personal authority. The transformation of human dignity into its modern conceptual usage was a gradual process, in which the work of Kant added an important dimension. Kant defines dignity as being a quality of intrinsic and absolute value.[103] Hence, in Kantian terms, human dignity signifies the intrinsic worth of the autonomous individual.

The recognition of the concept of human dignity as a fundamental right first occurred in international law through its inclusion in several international instruments, including the Preamble to the Charter of the United Nations[104] and the Universal Declaration of Human Rights.[105] Human dignity remains a malleable concept that is not amenable to precise definition as a legal term.[106] Although English law has yet to develop the concept of human dignity in a substantial way, the concept may well assume an increasingly elevated role following the incor-

[96] W Blackstone, *Commentaries on the Laws of England* (Oxford, 9th edn, 1783) vol I, 129.
[97] *Airedale NHS Trust v Bland* [1993] AC 789 (HL) 891 (Lord Mustill).
[98] I Kennedy and A Grubb, *Medical Law* (London, Butterworths, 3rd edn, 2000) 575.
[99] J Feinberg, *The Moral Limits of the Criminal Law* (Oxford, OUP, 1986) vol 3, 27.
[100] D Feldman, 'Human Dignity as a Legal Value—Part II' [2000] *Public Law* 61, 67.
[101] *Rees v Darlington Memorial Hospital NHS Trust* [2003] UKHL 52; [2004] 1 AC 309 (HL) [123].
[102] P Garnsey, *Social Status and Legal Privilege in the Roman Empire* (Oxford, Clarendon Press, 1970) 224–25.
[103] I Kant, *Foundations of the Metaphysics of Morals* LW Beck (tr) (New Jersey, Prentice Hall, 2nd edn, 1989) 39.
[104] (Signed 26 June 1945) UNTS 993.
[105] (Adopted 10 December 1948 UNGA Res 217 A(III)).
[106] D Feldman, 'Human Dignity as a Legal Value—Part I' [1999] *Public Law* 682, 682.

poration of article 8 of the European Convention on Human Rights (ECHR) into English law. Munby J has observed that it is 'important to note that personal autonomy and dignity are both aspects of "private life" respect for which is guaranteed by article 8 of the Convention'.[107] Article 8 provides (*inter alia*) that each person has 'the right to respect for his private and family life'. The approach of the European Court of Human Rights (ECtHR) to article 8 is important due to the potential impact that the ECtHR's jurisprudence in relation to the ECHR may have on the development of English law.[108] In broad terms, the ECtHR has held that the right to respect for private and family life protects both physical and moral integrity.[109] Moral integrity is related to human dignity; the ECtHR recently noted that the protection of moral integrity is intended to ensure the development 'of the personality of each individual in his relations with other human beings'.[110] The ECtHR, therefore, considers dignity to be an important aspect of article 8.

Drawing on Kant, the international instruments and the ECtHR jurisprudence, 'human dignity' is defined here as the respect that the law accords to the intrinsic worth of an individual. 'Autonomy' is defined in this book as being the capacity (and entitlement) that individuals have to act independently. The relationship between autonomy and human dignity was expressed by Lord Millett in *McFarlane v Tayside Health Board*, and is adopted here, that autonomy is an aspect of the broader concept of human dignity.[111] Following this view, the protection of autonomy can also be seen as the protection of human dignity.

In the context of medical law, bodily integrity is the foundational principle that protects an individual's autonomy.[112] The doctrine of consent is the basic doctrine by which the law protects an individual's right to act independently, and the House of Lords has recognised the central role of consent in a number of leading medical law cases.[113] Even in cases raising difficult moral questions, the common law has reinforced a living person's right to autonomy. As Lord Steyn observed in *Chester v Afshar*:

> The starting point is that every individual of adult years and sound mind has a right to decide what may or may not be done with his or her body. Individuals have a right to make important medical decisions affecting their lives for themselves: they have the right to make decisions which doctors regard as ill-advised.[114]

[107] *R (Burke) v General Medical Council* [2004] EWHC 1879; [2005] QB 424, [59]. Although the Court of Appeal upheld an appeal from this decision the court did not comment on Munby J's observations concerning autonomy at [59]: *R (Burke) v General Medical Council* [2005] EWCA Civ 1003; [2006] QB 273.

[108] Section 2(1)(a) of the Human Rights Act 1998 (UK) requires English courts to take relevant ECtHR jurisprudence into account when determining any question arising in connection with an ECHR right.

[109] *Costello–Roberts v United Kingdom* (1995) 19 EHRR 112, [49].

[110] *Case of Von Hannover v Germany* (2005) 40 EHRR 1, [50]; *Botta v Italy* (1998) 26 EHRR 241, [32].

[111] *McFarlane v Tayside Health Board* [2000] 2 AC 59 (HL) 114 (Lord Millett).

[112] *Re F (Mental Patient: Sterilisation)* [1990] 2 AC 1 (HL) 72 (Lord Goff).

[113] *Chester v Afshar* [2004] UKHL 41; [2005] 1 AC 134 (HL); *Airedale NHS Trust v Bland* [1993] AC 789 (HL); *Re F (Mental Patient: Sterilisation)* [1990] 2 AC 1 (HL).

[114] *Chester*, ibid at [14].

Tort law is the main mechanism by which English law protects an individual's right to bodily integrity. Interference (or threatened interference) with another person's bodily integrity without that person's consent may amount to trespass to the person or negligence. Trespass is not a cause of action but a form of action,[115] and includes assault and battery.[116] The tort of battery is the most relevant for the purposes of this book, and is defined as a wrong 'which is committed by intentionally bringing about a harmful or offensive contact with another person's body'.[117] In contrast, assault is an action that requires an immediate intention to commit a battery and the action is parasitic on the tort of battery.[118]

Battery is a tort that is actionable *per se*. Accordingly, it is not necessary to establish that physical injury has in fact resulted. As Robert Goff LJ observed in *Collins v Wilcock*,[119] '[t]he fundamental principle, plain and incontestable, is that every person's body is inviolate. It has long been established that any touching of another person, however slight, may amount to a battery'.[120] Thus, any intentional touching of a person is unlawful and may amount to the tort of battery unless justified by consent or other means. In the context of medical law, English courts[121] often invoke the statement of Cardozo J in *Schloendorff v Society of New York Hospital*:

> Every human being of adult years and sound mind has a right to determine what shall be done with his own body; and a surgeon who performs an operation without his patient's consent commits an assault, for which he is liable in damages.[122]

Examination of English case law reveals only a limited number of battery cases in the context of invasions of a person's body.[123] English courts have demonstrated a marked reluctance to expand the tort of battery in medical cases.[124] Where the patient is informed in broad terms of the procedure—and gives consent—no cause of action lies in battery.[125] Rather, English law has developed the tort of negligence in cases concerning the non-disclosure of information. This shift coincided with the emergence of negligence in the twentieth century as the dominant tort in private law.[126]

115 *Wainwright v Home Office* [2003] UKHL 53; [2004] 2 AC 406 (HL) [8] (Lord Hoffmann).

116 FA Trinidade, 'Intentional Torts: Some Thoughts on Assault and Battery' (1982) 2 *Oxford Journal of Legal Studies* 211, 211.

117 J Fleming, *The Law of Torts* (Sydney, LBC, 9th edn, 1998) 29.

118 *Wainwright v Home Office* [2001] EWCA Civ 2081; [2002] QB 1334, [68] (Buxton LJ).

119 [1984] 1 WLR 1172 (CA).

120 *Collins v Wilcock* [1984] 1 WLR 1172 (CA) 1177.

121 See, for example: *Re F (Mental Patient: Sterilisation)* [1990] 2 AC 1 (HL) 73 (Lord Goff).

122 211 NY 125 (NY CA 1914) 126.

123 See, for example: *Michael v Molesworth* [1950] 2 *British Medical Journal* 171 (different doctor performed the operation).

124 *Sidaway v Board of Governors of the Bethlem Royal and the Maudsley Hospital* [1985] AC 871 (HL) 882 (Lord Scarman).

125 *Freeman v Home Office* [1984] 1 QB 524, 537 (McCowan J); A Grubb, 'Consent to Treatment: The Competent Patient' in A Grubb (ed), *Principles of Medical Law* (Oxford, OUP, 2nd edn, 2004) [3.95].

126 T Weir, 'The Staggering March of Negligence' in P Cane and J Stapleton (eds), *The Law of Obligations* (Oxford, Clarendon Press, 1998) 98.

While the tort of battery focuses on the knowledge and understanding of the patient, the tort of negligence focuses on the conduct of the doctor.[127] It is well established in English law that a doctor owes a duty of care to their patient to act with reasonable care.[128] English law also requires doctors to obtain valid consent to perform lawful acts. A doctor must ensure that a patient receives sufficient information concerning the risks that any given procedure carries with it. As a result, the tort of negligence has assumed a central role in protecting an individual's bodily integrity. *Sidaway v Board of Governors of the Bethlem Royal and the Maudsley Hospital*[129] remains the leading English authority. The majority in *Sidaway* reaffirmed that a doctor must pass on to the patient such information as is thought appropriate according to a respected body of medical opinion.[130] Although criticised by some commentators,[131] *Sidaway* nonetheless represents the current state of English law,[132] and its correctness is not challenged here.

The criminal law also protects bodily integrity through the crimes of murder, manslaughter, battery, assault and rape. A person commits the crime of battery by intentionally or recklessly touching another person without the consent of the person and without lawful excuse.[133] English law applies Blackstone's observation:

> [T]he law cannot draw the line between different degrees of violence, and therefore totally prohibits the first and lowest stage of it; every man's person being sacred, and no other having a right to meddle with it, in any the slightest manner.[134]

As a result, unlawful bodily contact encompasses touching of any kind,[135] including, for example, the act of spitting.[136] In a statutory context, unlawful contact may also constitute the crime of assault (as amounting to wounding or causing grievous bodily harm) under sections 18, 20 and 47 of the Offences Against the Person Act 1861. The criminal law, therefore, reinforces tort law protection of the right to bodily integrity and can be contrasted with the lack of protection the law currently provides in relation to separated biological materials.

[127] For a discussion of the relationship between the two torts, see: TK Feng, 'Failure of Medical Advice: Trespass or Negligence' (1987) 7 *Legal Studies* 149, 150–55.

[128] *Bolitho v City & Hackney Health Authority* [1998] AC 232 (HL).

[129] [1985] AC 871 (HL).

[130] *Sidaway*, above n 124, at 893–94 (Lord Diplock), 900 (Lord Keith), 895 (Lord Bridge), 904–05 (Lord Templeman).

[131] Grubb, above n 125, at [3.145].

[132] *Chester*, above n 113, at [51] (Lord Hope).

[133] Offences Against the Person Act 1861, ss 18, 20 and 47; *Re Rolfe* (1952) 36 Cr App Rep 4, 5.

[134] Blackstone, above n 96, at vol III, 120.

[135] *Cole v Turner* (1704) 6 Mod 149, 149; 87 ER 907, 907.

[136] *Smith* (1866) F&F 1066, 1067; 176 ER 910, 910.

C Structure

This book has two parts. Part I (comprising chapters two, three and four) explores how English, Australian and US jurisdictions deal with legal questions concerning bodily materials separated from the human body.[137] Comparative analysis of these jurisdictions is valuable and necessary to develop an understanding of the legal principles that have underpinned the approaches of leading common law countries. The analysis reveals a muddled jurisprudence among the common law and statutory regimes. The muddle stems from a failure to identify clearly the underlying legal principles applicable to separated bodily materials. Without a clear set of guiding principles, the law encounters difficulties when faced with novel issues. It is in this state that the law currently finds itself with respect to separated bodily materials.

Part II (comprising chapters five, six and seven) analyses the future development of the law with respect to separated bodily materials. It builds a more rational structure for resolving questions concerning biological materials removed from living persons and dead bodies. Analysis is divided into two sections: first, chapters five and six consider property rights; secondly, chapter seven examines non-proprietary interests.

Chapters five and six analyse the role of the law of property, and particularly the circumstances in which property rights to biological materials separated from a living person can arise. Property rights enable both control over, and commercial participation in, physical entities. Even though English law has demonstrated a continuing reluctance to recognise personal property rights in separated biological materials, it is argued that the law can develop to accommodate such property rights.

In delimiting the existence and scope of rights to biological materials, two questions must be addressed: (1) how are biological materials transformed into 'things' capable of being the subject of property rights? and (2) what circumstances create property rights in biological materials? The book focuses on the second question, and argues that three circumstances explain the creation of property rights in separated biological materials. Property rights can be created on the basis of: (1) work or skill being performed on biological materials; (2) the detachment of biological materials; and (3) the detachment of biological materials coupled with an intention to use those materials as property.

Chapters five and six explore which of these three circumstances provides the most compelling basis for explaining the creation of property rights in terms of legal principle. It is argued that option (2) provides the most logical explanation and supports the creation and allocation of property rights to the individual source from whom the biological materials were removed. In addition, chapter six

[137] 'Bodily materials' is used as a general term in this book to include biological materials, gametes, hair clippings and waste products.

applies property principles in the practical context of scientific and medical research.

Chapter seven examines how the law of wrongs might also protect a source's non-proprietary interests of autonomy and freedom from emotional distress. In contrast to proprietary protection, the law of wrongs provides for control of physical entities without commercial participation. Chapter seven first considers whether English law should develop so as to protect an individual's right to autonomy and freedom from emotional distress in respect of separated biological materials. Causes of action that protect an individual's autonomy include breach of the duty of confidence and the torts of battery and negligence. The chapter then examines whether English law should recognise a separate tort of non-consensual use of separated biological materials.

Ultimately, this book presents a coherent structure for the legal principles that can apply to biological materials separated from living persons and dead bodies. In the larger picture of private law, it explores the external boundaries of personal property law and the law of tort, as well as the interaction between them.

Part I

2

Dead Body

Questions concerning the legal status of the human body first arose in the common law when the practice of grave-robbing began in the late seventeenth and early eighteenth centuries. During this period, corpses acquired monetary value as their usefulness in medical training was recognised.[1] The legal status of cadavers, and rights to them, are not simply issues of historical significance; they have gained prominence in the UK following a number of high profile inquiries into post-mortem practices at the Bristol Royal Infirmary, the Royal Liverpool Children's Hospital (Alder Hey) and other NHS hospitals during the late 1980s and early 1990s.[2] These inquiries focused on the removal and retention of biological materials (including organs) following stillbirths and abortions. They highlighted the unsatisfactory development of the common law with respect to rights concerning human bodies and separated biological materials.[3] Three points are developed in this chapter. First, the chapter traces the development and influence of the 'no property' principle in the context of cadavers. Secondly, it analyses the circumstances in which the law has recognised property rights in biological materials separated from dead bodies. Finally, it examines non-proprietary protections employed by the law.

A No Property Principle

The starting point for any analysis of the common law's protection of buried and unburied bodies is the line of English authority that establishes the principle that there is 'no property' in human corpses (the 'no property principle'). There is

[1] R Richardson, *Death, Dissection and the Destitute* (London, Phoenix Press, 2001) 52.

[2] Bristol Royal Infirmary Inquiry, *Report of the Royal Liverpool Children's Inquiry* (2001) (available at http://www.bristol-inquiry.org.uk/index.htm, accessed on 1 June 2007); Redfern Inquiry into the Royal Liverpool Children's NHS Trust–Alder Hey (2001) (available at http://www.rlcinquiry.org.uk, accessed on 1 June 2007); Independent Review Group, *Retention of Organs at Post-Mortem in Scotland* (2001) (available at http://www.show.scot.nhs.uk, accessed on 1 June 2007).

[3] Advice from the Chief Medical Officer, *The Removal, Retention and use of Human Organs and Tissue from Post-mortem Examination* (Department of Health, London, 2001); Chief Medical Officer, *Report of a Census of Organs and Tissues Retained by Pathology Services in England* (Department of Health, London, 2001). These documents are available at: http://www.doh.gov.uk (accessed on 1 June 2007).

similar authority in civil law jurisdictions.[4] Despite the antiquity of the 'no property' principle, analysis of the early English authorities suggests that early judicial support for the principle primarily took the form of *obiter dicta*. Further, it has been suggested that the rule itself arose from deficient reporting.[5] In England, the earliest common law authority is *Haynes's Case*,[6] where Haynes was charged with petty larceny and felonious taking in respect of certain winding sheets. It remains questionable whether this decision supports the 'no property' principle. The significance of the case lay in its determination of the person in whose name the charges should be laid. The court observed that the corpse did not have property rights in the sheets. This observation seems to have been misunderstood by later commentators as meaning that a corpse itself is not capable of being property.[7] Blackstone, for example, states in his *Commentaries on the Laws of England*, that: 'stealing the corpse itself, which has no owner, though a matter of great indecency, is no felony, unless some of the gravecloths be stolen with it'.[8]

If *Haynes's Case* is not authority for the 'no property' principle, the earliest appearance of the principle may have been in Sir Edward Coke's *The Institutes of the Laws of England*.[9] Coke claims that the 'no property' principle is self-evident through a simple exercise in etymology: a corpse cannot constitute property because the word 'cadaver' is an acronym for the Latin phrase *caro data vermibus*, meaning 'flesh given to worms'. Coke's reasoning, too, is questionable. Indeed, his research into the origins of the word cadaver appears to have been flawed. The *Oxford English Dictionary* notes that it was introduced into the English language as an adoption of the French version *cadaver*. This, in turn, was derived from the verb *cad-ere*, meaning 'to fall'.[10]

Despite these weak foundations, *Exelby v Handyside (Handyside's Case)*[11] gave explicit judicial recognition to the 'no property' principle. That case was never officially reported and first appeared in *East's Pleas of the Crown* (1803). *Handyside's Case* concerned an action in trover brought against a doctor for the bodies of two children joined by a birth defect. East reports that Willes CJ dismissed the action and held that it would not lie because no person has any property in corpses.[12] However, historical research has established that the case was settled while the jury was out.[13] As Magnusson suggests, this may explain why

[4] In Scotland, for example, see: *McGruer, Petr* (1899) 15 Sh Ct Rep 38, 38 (Scotland).

[5] P Matthews, 'Whose Body? People as Property' (1983) 36 *Current Legal Problems* 193, 197–200; RS Magnusson, 'Proprietary Rights in Human Tissue' in NE Palmer and E McKendrick (eds), *Interests in Goods and Services* (London, LLP, 2nd edn, 1998) 10.

[6] (1614) 12 Co Rep 113; 77 ER 1389.

[7] Matthews, above n 5, at 197–98.

[8] W Blackstone, *Commentaries on the Laws of England* (Oxford, 9th edn, 1783) vol IV, 2.

[9] E Coke, *The Third Part of the Institutes of the Laws of England* (London, 4th edn, 1669) vol III, 203.

[10] JA Simpson and ESC Weiner (eds), *The Oxford English Dictionary* (Oxford, Clarendon Press, 2nd edn, 1989) vol II, 757.

[11] (1794) 2 East PC 652.

[12] *Handyside's Case* (1749) 2 East PC 652.

[13] Matthews, above n 5, at 208–10; R Hardiman, 'Toward the Right of Commerciality: Recognising Property Rights in the Commercial Value of Human Tissue' (1986) 34 *UCLA Law Review* 207, 225.

Sir John Willes CJ is reported as saying that the action *would* not lie rather than that it *did* not lie.[14]

On the basis of the 'no property' principle ecclesiastical courts had exclusive jurisdiction in matters relating to human corpses, including (but not limited to) the disposal of corpses by burial. Ecclesiastical jurisdiction was recognised in *R v Coleridge*,[15] where an application to the King's Bench was refused because the court considered that the mode of burial was a question for the ecclesiastical courts.[16] Because ecclesiastical courts prevented unwarranted interference with cadavers[17] there is little English legal academic writing on the topic.

By the middle of the nineteenth century, the common law had established and developed some jurisdiction over buried corpses. Common law jurisdiction evolved because the courts increasingly had to consider questions concerning corpses buried in unconsecrated ground not protected by ecclesiastical law.[18] The 'no property' principle became commonly invoked and its application meant that cadavers could neither be owned nor stolen.[19] Unauthorised exhumation was not, therefore, technically a crime of theft. Prosecution for a felony only occurred if a portion of the grave clothes or coffin was stolen with the body.[20] In response, the common law developed other means of protecting cadavers. *R v Lynn*[21] held that an unlawful disinterment was an offence that was 'cognizable in a Criminal Court, as being highly indecent and *contra bonos mores*'.[22] Later authorities supported the criminal offence.[23] Other common law crimes developed to cover unburied corpses, including preventing the burial of a dead body[24] and conspiring to prevent burial.[25] Statutory intervention also purported to regulate the use of corpses, but the statutory regime did not create property rights to cadavers.[26] English common law therefore started from the 'no property' principle and initially invoked the criminal law to provide protection.

14 Magnusson, above n 5, at 28.

15 (1819) 2 B & Ald 804; 106 ER 559.

16 *Coleridge*, ibid, at 805; 560.

17 R Burn, *Ecclesiastical Law* (Dublin, Sweet & Stevens, 9th edn, 1842) vol I, 271a.

18 Land becomes consecrated when the bishop of a diocese signs a document that separates an area of land and dedicates it to the services of Almighty God: *In re Blagdon Cemetery* [2002] 3 WLR 603 (Arches Ct of Cant) 607. Unconsecrated ground is not set apart and is not, therefore, within the ecclesiastical jurisdiction.

19 Despite these legal developments, a number of members of the community—anatomists, artists, physicians, surgeons and their suppliers—treated cadavers as commodities that could be bought and sold: Richardson, above n 1, at 71–72.

20 Blackstone, above n 8, at vol IV, 235. See, also: JB Bailey, *The Diary of a Resurrectionist* (London, Swan Sonnenschein & Co, 1986) 90.

21 (1788) 2 TR 733; 100 ER 394.

22 *Lynn*, ibid, at 734; 395.

23 *R v Sharpe* (1856–1857) Dears & Bell 160, 163; 169 ER 959, 960; *R v Price* (1884) 12 QB 247, 252; *Foster v Dodd* (1867) LR 3 QB 67, 77.

24 See unreported case *R v Young* (1784) 4 Wentworth Pleadings 219 referred to in *Lynn*, above n 21.

25 *R v Hunter* [1974] 3 WLR 374 (CA).

26 The original Anatomy Act 1832 2 & 3 Will 4, c 75 was passed as a reaction to the unregulated practice of body snatching. Under the Anatomy Act 1832, donations were permitted by persons in lawful possession of the corpse, or by the deceased prior to his or her death. The Anatomy Act 1832 was replaced by the Anatomy Act 1984.

In modern times, the 'no property' principle has been affirmed by the Court of Appeal in *R v Kelly*[27] and, more recently, in *Re Organ Retention Group Litigation.*[28] Both of these cases are analysed below. Moreover, the Human Tissue Act 2004 UK (HTA 2004) does not purport to alter the common law 'no property' principle.[29] The 'no property' principle is also recognised in Australia, the US and Canada. Although it is a matter of debate whether ecclesiastical law applies in Australia, Australian courts have generally deferred to the English 'no property' principle.[30] Similarly, although not influenced by ecclesiastical courts, American[31] and Canadian[32] courts have frequently invoked the 'no property' principle. Thus, the 'no property' principle has taken hold in England, Australia, the US and Canada despite conspicuously dubious historical origins.

B Property Rights

On the basis of the 'no property' principle, the general position at common law is that dead bodies and biological materials separated from them cannot be the subject of property rights. Nevertheless, exceptions to the general principle exist in English, Australian and US jurisprudence.

1 Work or Skill Exception

The work or skill exception, as originally formulated in the Australian case of *Doodeward v Spence*,[33] represents the most significant common law recognition of property rights to biological materials separated from dead bodies. In *Doodeward*, the High Court of Australia held that 'a human body, or a portion of a human body, is capable in law of becoming the subject of property'.[34] A doctor had preserved the corpse of a pair of stillborn conjoined twins in a jar. Following the doctor's death, the preserved corpse was sold to the plaintiff. It was then confiscated by a police inspector (the defendant) because the plaintiff was exhibiting the corpse in public. The plaintiff requested that the exhibit be returned, and, although the bottle and preservation liquid were returned, the defendant refused

[27] [1999] QB 621, 630.

[28] [2004] EWHC 644; [2005] QB 506.

[29] HC Standing Committee G, col 215 (3 February 2004) (Dr Ladyman). See, also: Department of Health, *Proposals for New Legislation on Human Organs and Tissues* (Department of Health, London, 2003) 2.

[30] *Burrows v Cramley* [2002] WASC 47 (WA SC) [15]; *AW v CW* [2002] NSWSC 301 (NSW SC) [15]; *Re Gray* [2001] 2 Qd R 35 (Qld SC) 138; *Roche v Douglas* (2000) 22 WAR 331 (WA SC) 338; *Doodeward v Spence* (1908) 6 CLR 406 (HCA) 414.

[31] *Enos v Snyder* 63 P 170 (Cal SC 1900) 171; *O'Donnell v Slack* 55 P 906 (Cal SC 1899) 907.

[32] *Edmonds v Armstrong Funeral Home* [1931] DLR 676 (Alb SC) 680.

[33] (1908) 6 CLR 406 (HCA).

[34] *Doodeward*, ibid at 414.

to give back the corpse. Consequently, the plaintiff sued for the return of his property and at first instance the District Court dismissed the matter. The NSW Supreme Court dismissed an appeal against the decision. It held that because a human body was not property it could not be the subject of a proprietary action.[35]

The High Court of Australia allowed an appeal. Delivering the leading judgment, Griffith CJ commenced by observing that an unburied corpse awaiting burial is *nullius in rebus*. However, his Honour considered that this principle did not mean that a corpse was 'incapable of becoming the subject of ownership'.[36] By analogy, the dead body of an animal *ferae naturae* is not the property of anyone and can be appropriated by the finder.[37] The question for consideration was whether the 'continued possession of a human corpse unburied is *in re ipsa* unlawful'.[38] Griffith CJ considered possession to be unlawful where it proved injurious to the public welfare. If possession was not injurious to the public welfare, Griffith CJ held that the law should protect such possession by appropriate remedies. After noting the collections of anatomical and pathological specimens maintained by scientific bodies, Griffith CJ found that such possession was not unlawful where preservation could 'afford valuable or interesting information or instruction'.[39] The property rights, in such cases, arose from the lawful exercise of work or skill:

> [A] human body or a portion of a human body is capable by law of becoming the subject of property. It is not necessary to give an exhaustive enumeration of the circumstances under which such a right may be acquired, but I entertain no doubt, that when a person has by the lawful exercise of work or skill so dealt with a human body or part of a human body in his lawful possession that it has acquired some attributes differentiating it from a mere corpse awaiting burial, he acquires a right to retain possession of it.[40]

Griffith CJ did not explain why (as a matter of legal principle) such work or skill creates property rights, and it is unclear whether the other judges supported this proposition. Barton J, the other majority judge, restricted his reasoning to stillborn infants, but did generally concur with the reasons of Griffith CJ.[41] Significantly, Barton J did not question the general rule that an unburied corpse is not the subject of property. On the facts, Barton J found that the foetus 'never existed independently of the physical attachment to its mother. It was never alive in the ordinary sense of human life'.[42] Barton J thus concluded that the law could not characterise a stillborn foetus as an unburied corpse.

Higgins J dissented, considering the threshold question to be whether property could exist in a dead human body. His Honour opined that, without property rights, there could be no action of detinue or trover. In this respect, Higgins J relied

35 *Doodeward v Spence* (1907) 24 NW (NSW) 189 (NSW SC) 190.
36 *Doodeward*, above n 33, at 411.
37 Ibid.
38 Ibid.
39 Ibid, at 414.
40 Ibid.
41 Ibid, at 417.
42 Ibid, at 416.

on the established line of English authority that no one can have property in another person.[43] In *Doodeward*, as the body was not wanted for burial, Higgins J concluded that it could not be protected.[44]

Nevertheless, *Doodeward* represents the starting point of the work or skill exception and warrants closer analysis. Three qualifications are discernible from Griffith CJ's judgment. The first is that the person claiming the property rights must have lawful possession of the body or body part. Secondly, that person must perform the work or skill on the body or body part. Thirdly, that person does not obtain an ownership right in the body or body part but a lesser possessory interest. A possessory interest gives the possessor the right to retain the body against all the world except the person who is entitled to bury the body.

The precise nature of the interest that an individual obtains to separated biological materials under the exception is not clear from Griffith CJ's judgment. The rights are variously referred to as 'a right to retain possession of it'[45] and 'a permanent right to possession'.[46] It was also noted that 'a human body, or a portion of a human body is capable by law of becoming the subject of property'.[47] On the basis of these statements, the right recognised in *Doodeward* might best be described as simply a right to possession.

Significantly, Griffith CJ did not state that the acquisition of different attributes is necessary for property rights to arise. He noted that it was not necessary to 'give an exhaustive enumeration of the circumstances under which such a [property] right may be acquired'.[48] This left open the possibility that property rights could be created without any physical change to the separated body part. More recent Australian authority has not considered the work or skill exception, but simply has noted its existence in *Doodeward*.[49]

Dobson v North Tyneside Health Authority[50] presented the first modern opportunity for the English common law to consider the 'no property' principle. In *Dobson*, the deceased's next-of-kin (the claimants) brought medical negligence proceedings against the North Tyneside Health Authority (the first defendant) for failing to diagnose brain tumours in the deceased, and the Newcastle Area Health Authority (the second defendant) for failing to preserve the brain of the deceased. The deceased had collapsed at work and was initially diagnosed with generalised epilepsy. She later became unwell and further investigations revealed two brain tumours, though she died before an operation could be performed. The coroner requested a post-mortem examination, and during the autopsy, the deceased's brain was removed and fixed in paraffin. The deceased's body (without the brain) was subsequently returned to her family for burial. The claimants sued the first defendant in

[43] *Doodeward*, above n 33, at 422.
[44] Ibid, at 424.
[45] Ibid, at 414.
[46] Ibid.
[47] Ibid.
[48] Ibid.
[49] *AW v CW* [2002] NSWSC 301 (NSW SC) [15].
[50] [1997] 1 WLR 596 (CA).

negligence for failing to detect the brain tumours. It is difficult to discern what cause of action was pleaded against the second defendants. The statement of claim against the second defendants alleged that they were 'gratuitous bailees' and were not entitled to destroy, lose or otherwise wrongfully interfere with the brain of the deceased pending further investigation. Although no cause of action was pleaded, Peter Gibson LJ concluded that the primary claim must have been for conversion.[51]

The case came before the Court of Appeal after the District Court struck out the claim against the second defendant. The leading judgment of Peter Gibson LJ (with whom Thorpe and Butler-Sloss LJJ agreed) is relevant for its discussion of the legal status of corpses generally, and of cadaveric specimens specifically. Peter Gibson LJ initially noted that there are two qualifications to the general principle that there is no property in a corpse. First, as discussed below, there is a right of possession for the purpose of burial or disposition of the body and, secondly, biological materials separated from a body may attain proprietary status following the application of human skill.[52]

In relation to the first qualification, Peter Gibson LJ observed that 'the executors, administrators and others charged by law with the duty of interring the body have a right to the custody and possession of it until it is properly buried'.[53] He held that the rule did not apply in *Dobson* because the administratrix was not appointed until after the deceased was buried.

Regarding the second qualification, the Court of Appeal accepted that a body part may be subject to rights of possession if a person lawfully in possession of it does work or exercises skill such that the body part acquires attributes differentiating it from a mere corpse. It is not apparent from Peter Gibson LJ's judgment whether he was referring to an ownership interest or simply 'rights to possession'. In acknowledging these rights in English law for the first time, Peter Gibson LJ accepted the classic statement of principle enunciated by Griffith CJ in *Doodeward*. On the facts Peter Gibson LJ found that the work or skill exception did not apply because the preservation of the brain was not a sufficient exercise of work or skill.[54]

The work or skill exception arose again for consideration in *R v Kelly*.[55] The question in *Kelly* was whether 40 human body parts taken by Kelly and Lindsay (the appellants) from the Royal College of Surgeons (the College) were capable of being stolen under the Theft Act 1968 UK. Kelly had persuaded Lindsay (a technician employed by the College) to remove the specimens for £400. There were between 35 and 40 specimens, all of which had been preserved or fixed by College staff involving hours and, in some cases, weeks of skilled work.[56] Kelly made use of the body parts in his work as an artist and used some of the parts to make casts, which were exhibited in an art gallery.

[51] Ibid, at 596, 602 (CA).
[52] Ibid, at 600.
[53] Ibid.
[54] Ibid, at 601.
[55] [1999] QB 621.
[56] Ibid, at 623.

The appellants were charged with theft under section 1 of the Theft Act. Section 1(1) defines 'theft' as the dishonest appropriation of 'property' belonging to another with the intention of permanently depriving the other of it. 'Property' is defined in section 4(1) to include real and personal property, including choses in action and other intangible property. Section 5(1) sets out that property is regarded as belonging to a person having possession or control of it, or having a proprietary right or interest in it.

At trial, the defence submitted that the charge should be dismissed because body parts were not property, and therefore were not capable of being stolen. The trial judge rejected this submission and ruled that a body part could be capable of becoming property, in accordance with *Doodeward.* Kelly and Lindsay were both convicted. They appealed, arguing that the jury's verdict was unsafe on two grounds. First, they argued that the body parts were not property and could not be stolen. Secondly, they claimed that the College was not lawfully in possession of the body parts by virtue of provisions in the Anatomy Act 1932 UK. Their appeal was dismissed. Rose LJ delivered the judgment of the Court of Appeal (Ognall and Sullivan LJJ agreeing) and observed that, despite the questionable origins of the 'no property' principle, only Parliament could change the law because of its entrenched nature.[57] Rose LJ held that since *R v Sharpe*, 'it has been express or implicit in all the subsequent authorities and writings to which we have been referred that a corpse or part of it cannot be stolen'.[58] Rose LJ found, nevertheless, that parts of a corpse are capable of becoming property within section 4 of the Theft Act if they have 'acquired different attributes' through the application of skill (such as dissection or preservation techniques) for exhibition or teaching purposes.[59] He cited *Doodeward* in support and observed that the Court of Appeal did not dissent from that case in *Dobson.*

Rose LJ did note the flexible nature of the common law and predicted that a future court might hold that there could be property in unaltered body parts in different circumstances. As he observed:

> [T]he common law does not stand still. It may be that if, on some future occasion, the question arises, the courts will hold that human body parts are capable of being property for the purposes of s 4, even without the acquisition of different attributes, if they have a use or significance beyond their mere existence. This may be so if, for example, they are intended for use in an organ transplant operation, for the extraction of DNA or, for that matter, as an exhibit in a trial.[60]

This view has been criticised, with Grubb suggesting that this would 'in effect rewrite the "no property" rule itself such that it became: "there is property in corpses (excluding excised body parts)"'.[61] Rose LJ's *obiter dictum* is significant as

[57] Ibid, at 630–31.
[58] Ibid, at 631.
[59] Ibid.
[60] Ibid.
[61] A Grubb, 'I, Me, Mine: Bodies, Parts and Property and Parts' (1998) 3 *Medical Law International* 299, 312.

it suggests a different basis (which does not require the application of work or skill) for the creation of property rights. Surprisingly, it has been the subject of little academic scrutiny. According to Rose LJ, body parts are capable of a proprietary characterisation where they have a use beyond their mere existence. Rose LJ's suggestion is examined in greater detail in chapter six.

Although no Australian decision has considered *Doodeward* in any detail, a recent decision—*Leeburn v Derndorfer*—provided a further context for the application of the work or skill exception.[62] In this case the executors were the three adult children of the deceased and they were unable to agree as to the disposal of their father's ashes. The deceased had been cremated pursuant to his will but had failed to provide further instructions as to the disposal of the ashes. Two executors (the daughters) wanted the ashes interred at a cemetery and the third executor (the son) wanted the ashes divided so that each executor received a third. The daughters collected the ashes and had them interred in a sealed container at a cemetery. On discovering this action, the son brought proceedings seeking an order that the ashes be disinterred and that he be given one-third of them for disposal as he thought fit. Ultimately, Bryne J decided the question on the basis of 'practicalities' and concluded that the ashes should remain at the cemetery, where they had been for four years. Bryne J did address the potential application of the work or skill exception in the context of ashes and he observed:

> Moreover, so long as they are not dispersed or otherwise lose their physical character as ashes, they may be owned and possessed. To my mind, therefore, it is apt to characterise the legal status of the ashes as similar to that of the preserved body in *Doodeward v Spence*. In this way the application of fire to the cremated body is to be seen as the application to it of work or skill which has transformed it from flesh and blood to ashes, from corruptible material to material which is less so.[63]

It is not clear whether there are any limitations in the context of ashes because, as Bryne J noted, a qualification may arise because ashes do represent the remains of a human being and should therefore be treated with respect and reverence.[64] One commentator has supported the application of property principles in the context of the disposal of human ashes.[65] If a property right does exist, the next question is in whom does the property right vest? Importantly, because ashes are not part of the estate they do not pass under the will to the executors.[66] Bryne J considered instead that the interest of the executors in the ashes is that of trustee. The executors held the ashes as trustees for the 'purpose of disposing or dealing with them in a way that seem to them to be appropriate having regard to any direction of the deceased in the will or otherwise and having regard to the claims of the relatives'.[67] This conclusion is inconsistent with a proper application of the work

[62] [2004] VSC 172.
[63] *Leeburn*, ibid at [27].
[64] Ibid.
[65] M Groves, 'The Disposal of Human Ashes' (2005) 12 *Journal of Legal Medicine* 267, 272.
[66] *Leeburn*, above n 62, at [28].
[67] Ibid.

or skill exception, which would vest the property right in the person or body that performs the cremating.[68] On the basis of the work or skill exception the executors could not hold the ashes as trustees as Bryne J concludes. Bryne J's analysis, thus, appears inconsistent with the work or skill exception as developed in *Kelly*.

Re Organ Retention[69] is the most recent common law case to examine the work or skill exception. It raises several important issues in the context of separated biological materials, including property rights to cadaveric specimens and the potential for a tort of wrongful interference.[70] While these issues are addressed later in this chapter, it is useful to present the background of the case. The *Re Organ Retention* litigation arose following inquiries at the British Royal Infirmary and Alder Hey Hospital, which exposed the long-term practice of retaining biological materials (including organs) from children who had died. Proceedings were commenced by aggrieved parents who were collectively known as the National Organ Group Litigation (NOGL). At the time of the decision there were 2,140 claims on the NOGL register.[71] Three lead actions were brought by Karen and David Harris, Susan Carpenter and Denise Shorter. The lead claimants alleged that their children's biological materials had been removed, retained and disposed of without their knowledge. To assist in the resolution of the other registered claims, the parties also requested the High Court to decide four preliminary issues concerning the use of removed organs for research, based on an agreed factual matrix.

The central question in *Re Organ Retention* was whether the removal and retention of organs from children during post-mortems was unlawful. The claimants alleged that the defendants: (1) committed the tort of wrongful interference; (2) were negligent, causing psychiatric injury; and (3) breached the Human Rights Act 1998 UK. The claimants sought general and special damages arising from the psychiatric injury that they suffered. They also claimed exemplary and aggravated damages. Gage J concluded that no tort of wrongful interference at law had been committed, but a negligence claim was prima facie available against the hospitals. Only in Shorter's case did Gage J make an award for psychiatric injuries suffered.[72] In Harris's case, Gage J found that the psychiatric injury was not reasonably foreseeable,[73] and, in Carpenter's case, the discovery that the child's organs had been retained did not materially contribute to Carpenter's psychiatric illness.[74]

[68] *Kelly*, above n 55, at 631.

[69] *Re Organ Retention Group Litigation* [2004] EWHC 644; [2005] QB 506.

[70] *Stevens v Yorkhill NHS Trust and Southern General University Hospitals NHS Trust* [2006] CSOH 143; 2006 SLT 889 (Outer House) was a similar test case, which concerned the retention of children's organs at post mortem in Scotland and is discussed below at nn 235–57 and accompanying text.

[71] *Re Organ Retention*, above n 69, at [3]. A large number of the claims arsing out of the Alder Hey Hospital were pursued under a separate Group Litigation Order and settled out of court in 2003.

[72] *Re Organ Retention*, above n 69, at [267].

[73] Ibid, at [259].

[74] Ibid, at [278].

In respect of the tort of wrongful interference, the defendants argued that property rights to separated biological materials vested in them. Their submission was based on two propositions: first, that the removal of biological materials was authorised under statute and common law as part of post-mortem procedures; and secondly, that the application of work or skill on biological materials had created property rights that were allocated to the defendants. Each proposition requires separate consideration.

The question of whether the removal of the biological materials was lawful hinged on what may lawfully be done in the context of post-mortems. There are two forms of post-mortems: hospital post-mortems (HPMs) (relevant in the lead cases of Harris and Shorter) and the coroner's post-mortem (CPM) (relevant in the lead case of Carpenter). An HPM is conducted so that relatives may know the cause of death. Section 1 of the Human Tissue Act 1961 (HTA 1961)[75] (since repealed by the HTA 2004) required that before such a procedure could be carried out, the persons responsible must reasonably have believed that the deceased's relatives would not object. At common law, the hospital's possession of the body remained lawful throughout the post-mortem process, so long as the parents had not objected in accordance with the HTA 1961.[76]

In contrast to the hospital HPM, coroners have a statutory duty to conduct an inquest whenever they have reasonable cause to suspect that the deceased may have died a violent or unnatural death.[77] If the coroner directs a post-mortem examination, they must appoint a pathologist to carry it out. Gage J found that the HPM and the CPM organ removals were lawful. If the possession of the body and removal of the biological materials is lawful, the issue then becomes what rights subsist in the separated biological materials following the post-mortem. A properly executed post-mortem will inevitably involve the retention of material.[78] In *Re Organ Retention*, Gage J found that the removal of the biological materials was lawful and the remaining question thus became what rights existed following their removal. The statutory regimes concerning post-mortems do not address this issue. For example, under the applicable rules the coroner can preserve materials that bear on the cause of death for such a period as they think fit.[79] It is not settled at English law how long the coroner's authority over the body or any of its parts endures, or what happens once this authority concludes.[80]

The defendants contended that property rights to the removed biological materials were allocated to them at common law on the basis of the work or skill exception identified in *Kelly*. This submission was accepted by Gage J. Three aspects of the defendants' argument are significant: (1) the nature of the interest

[75] 9 & 10 Eliz 2.

[76] HTA 1961, s 1(7).

[77] Coroners Act 1988 UK, s 8(1).

[78] Closing Submissions (Defendants) *AB v Leeds Teaching Hospital NHS Trust & Cardiff & Vale NHS Trust* (High Court No HQ 0101462 12 February 2004) (copy on file with author) [213].

[79] Coroner's Rules 1984, r 9.

[80] *Re Organ Retention*, above n 69 at [23]. See, also: C Dorries, *Coroners' Courts* (Oxford, OUP, 2nd edn, 2004) 121.

conferred; (2) the degree of skill necessary for the work or skill exception to attach; and (3) in whom the property right was vested. In relation to the first aspect, the defendants submitted that it was not necessary for the court to consider whether the possessory rights amounted to an ownership interest.[81] It was argued that the possessor must at least have a right to permanent possession, and that the right to possession may differ little in substance from the right of an owner.[82] Gage J accepted this contention.

The second substantive question concerned the degree of work or skill necessary for the exception to apply. In *Re Organ Retention*, the defendants submitted that there was a sufficient exercise of skill through the pathological processes of selection, preservation and dissection of biological materials and the creation of tissue blocks.[83] The defendants noted that '[t]he fixation process irreversibly changes the tissue at the molecular level via the operation of the formaldehyde, and enables subsequent staining reactions (during the creation of slides) to develop in a predictable way'.[84] It is not clear, as a matter of scientific principle, that the defendants' analysis is correct. The defendants' use of the phrase 'irreversibly changes' implies that there had been an exchange of material at the molecular level. In fact, the fixation process preserves the cells essentially in their original size, shape, and position within the tissue by cross-linking its proteins to each other and those of its neighbouring cells. The fixed biological materials retain the original protein constituents, and therefore the materials maintain their molecular shape as they existed in their original form.[85] Although it is unnecessary to engage in detailed scientific analysis, an understanding of the relevant scientific principles is important. This book returns, in Part II, to analyse the applicable scientific principles and biological processes. For present purposes, the important factor is that in *Re Organ Retention* some qualifying work or skill was performed by technicians on the separated biological materials.

Gage J accepted that the work or skill exception had long been an established principle in English law. He noted that 'part of a body may acquire the character of property' where that part 'has been the subject of skill such as dissection or preservation techniques'.[86] Gage J considered in this case that sufficient work and skill had been employed:

> The evidence in the lead cases shows that to dissect and fix an organ from a child's body requires work and a great deal of skill, the more so in the case of a very small baby such as Rosina Harris. The subsequent production of blocks and slides is also a skilful operation requiring work and expertise of trained scientists.[87]

[81] Closing Submissions (Defendants), above n 78, at [233].
[82] Ibid, at [237].
[83] Ibid, at [247].
[84] Ibid, at [247].
[85] See: http://www.ncbi.nlm.nih.gov/entrez (accessed on 1 June 2007).
[86] *Re Organ Retention*, above n 69, at [148].
[87] Ibid.

In this way, *Re Organ Retention* extends the common law application of the work or skill exception in *Kelly* to tissue blocks, a type of biological material. Gage J then applied the work or skill principle and found:

> [F]ollowing the post-mortem, the hospital acquired proprietary and possessory rights to the organs. Once the pathologists had by their work and skill removed the organs and prepared blocks and slides for histological examination those organs and the blocks and slides came within the *Doodeward* exception.[88]

It is unclear from this passage whether Gage J considered that an ownership interest in the biological materials was allocated to the hospitals. In the statement above, he referred to 'proprietary and possessory rights'. Later he noted the 'rights to possession'[89] and commented that the hospital had 'lawful possession of the organs'.[90] Gage J was not required to resolve whether the hospital had an ownership interest because he accepted the defendants' submission that a right to possession was sufficient.

The third issue in *Re Organ Retention* concerned which person was to be allocated the property rights created under the work or skill exception. At one point Gage J accepted the statement in *Clerk and Lindsell on Torts*,[91] that the property rights vest in the person who alters the biological materials on the basis that they are the first possessor.[92] Later, however, he suggested that in some circumstances conversion might lie:

> It, on the other hand, a parent or parents when consenting to a post-mortem specifically ask for the return of an organ I can see in certain circumstances it might be arguable that a cause of action based on conversion exists, if conversion is what is being alleged in this group action.[93]

This statement is surprising. As Gage J recognised, the starting point at common law is the 'no property' principle. It is not possible to possess a dead body as 'property' at law. Dead bodies are not things capable of ownership and so are not protected by the tort of conversion. Yet in the passage quoted above, Gage J appears to endorse a possible application of conversion. There are three possible interpretations of Gage J's observation. The first is that his statement was simply an oversight and was not intended to deviate from the 'no property' principle. This interpretation is unpersuasive as it must be assumed that Gage J's express statement was deliberate.

Secondly, Gage J may have considered that parents could be allocated property rights at the time their children's biological materials are removed, such that the parents might request their return. Two factors militate against this conclusion.

[88] Ibid, at [257].
[89] Ibid, at [160].
[90] Ibid, at [257].
[91] A Tettenborn, 'Wrongful Interference with Goods' in AM Dugdale (ed), *Clerk and Lindsell on Torts* (London, Sweet & Maxwell, 19th edn, 2006) 1024.
[92] *Re Organ Retention*, above n 69, at [156].
[93] Ibid, at [161].

Gage J explicitly recognised that a dead body is not a thing at common law.[94] Further, he held that separated biological materials are capable of becoming things at common law only where they have acquired different attributes.[95] Gage J's statement could not properly be based on an assertion that the parents gain property rights at the time biological materials are separated because, at the time of separation, biological materials have not acquired different attributes through the application of work or skill.

A third and sounder interpretation of Gage J's statement is that biological materials are recognised as property at common law only when they have acquired different attributes. For example, an organ acquires different attributes when the pathologist fixes and hardens it. The difficulty with this third interpretation lies in explaining how it is that the *parents*—and not the pathologists—are entitled to protection of property rights through the tort of conversion. Gage J had accepted earlier the conclusion of *Clerk and Lindsell* that in such circumstances the property rights vest in the person who performed the work.[96] This problem may be illustrated by constructing a scenario based on Gage J's *dictum*. Assume that parents permit the removal of an organ from the body of their child during a post-mortem and ask for the organ to be returned to them once it has been used. The pathologist then takes the organ and proceeds to fix and harden it. This work is sufficient to fall within the work or skill exception. Under the third interpretation of Gage J's approach, property rights would not be allocated to the pathologist but instead to the parents, because they asked for the organ to be returned. If the property rights were allocated to the pathologist, then the parents would have no cause of action in conversion. Arguably, a form of bailment would exist such that the pathologist would hold the organ as a bailee for the parents as bailors. An action in conversion might lie if the pathologist failed to return the organ or acted inconsistently with the parents' property rights.

If the third interpretation of Gage J's statement is correct, it is still not clear why, simply by asking for the return of the organ, the parents are allocated property rights. *Kelly* establishes that the work or skill exception allocates property rights to the person who performs the work or skill. Gage J's application of this principle is inconsistent with the exception. Gage J's approach is incongruous and, viewed in the light of the above scenario, demonstrates the common law's ongoing failure to identify the correct legal principle underlying the work or skill exception. This failure is regrettable because the exception has assumed broader application under the HTA 2004, which adopts the exception and extends it to cover biological materials separated from a *living* person.[97] Chapter four critically examines this statutory extension.

Thus, the nature and contours of the common law exception are not clearly established. Difficult questions are yet to be addressed adequately by the courts

94 *Re Organ Retention*, above n 69, at [135].
95 Ibid, at [155].
96 Ibid, at [156].
97 HTA 2004, s 32(9)(c).

and five key questions remain unanswered. First, must there be an intention to create a novel item, or is it sufficient that work or skill of some sort is carried out on separated biological materials? *Kelly* suggests that an intention to create a novel item is required because the court in that case required that the body part should be altered for a special purpose.[98] Secondly, will work alone suffice or must there be an exercise of skill? *Kelly* referred to skill.[99] In *Re Organ Litigation*, Gage J referred to skill by itself and then to 'work and skill'.[100] Moreover, if an application of skill is required to enliven the exception, it is not clear what standard of skill is required. In *Kelly*, Rose LJ talked of dissection and preservation techniques, but such formulations do not provide a standard capable of general application.[101] In *Dobson*, Peter Gibson LJ suggested that the fixing of a body part in paraffin would not be sufficient to activate the exception.[102]

The third unresolved question concerns the nature of property rights recognised by the common law work or skill exception. Is it simply a right to possession, or instead does it confer property rights that the law recognises with respect to normal chattels? *Clerk and Lindsell on Torts* suggests that separated biological materials seem to be 'the subject of property in the ordinary way'.[103] In *Doodeward*, Griffith CJ's judgment referred to 'a right to retain possession of [the body]'[104] and 'a permanent right to possession'.[105] *Re Organ Retention* mentions proprietary and possessory rights.[106] *Kelly* did not address this question.

The fourth question is to whom should the property rights be allocated? There are several possibilities, including the source (in the case of a living person), the family of the deceased (in the case of a dead body), the person who alters the biological materials, or the employer of the person who alters the biological materials. *Clerk and Lindsell on Torts* states that presumably the property right will vest in the person who alters the body part on the basis that they are the first possessor.[107] *Kelly* held that property rights should be allocated to the person (or employer) who performed the work or skill. *Leeburn* suggested that property rights over ashes can vest in an executor as a trustee. In *Re Organ Retention*, even though the statement in *Clerk and Lindsell on Torts*[108] was approved, Gage J appears to suggest that in certain circumstances property rights could be acquired by (or transferred to) a family member.[109]

The final question is whether the common law exception is confined to biological materials separated from dead bodies, or does it also apply to biological

98 Grubb, above n 61, at 250.
99 *Kelly*, above n 55, at 631.
100 *Re Organ Retention*, above n 69, at [257].
101 *Kelly*, above n 55, at 631.
102 *Dobson*, above n 50, at 601.
103 Tettenborn, above n 91, at 1024.
104 *Doodeward*, above n 33, at 414.
105 Ibid.
106 *Re Organ Retention*, above n 69, at [161].
107 Tettenborn, above n 91, at 1024.
108 *Re Organ Retention*, above n 69, at [156].
109 Ibid, at [161].

materials separated from living persons? In *Kelly*, Rose LJ simply referred to 'human body parts'.[110] At one point in *Re Organ Retention*, Gage J also indicated that 'part of a body may acquire the character of property'.[111] Nevertheless, both these comments were made in the context of biological materials removed from dead bodies and it is likely that the principle at common law is currently so confined. Yet, there is no reason in principle why the common law exception should be limited to bodily materials separated from cadavers.

These inconsistencies and unresolved questions demonstrate that the legal principle underpinning the work or skill exception has not been adequately identified or articulated. Without a clear underlying principle the exception cannot develop in a coherent manner. Academic commentators have suggested that the exception represents an invocation of the 'first possession' principle.[112] Part II of this book rejects that explanation. Chapter five argues that the specification doctrine provides a more logical explanation of the work or skill exception.[113] Specification, as derived from the Roman law principle of *specificatio*, is the principle that applies where a new thing is brought into existence.[114] Chapter five considers the respective strength of the first possession and specification doctrines as contending legal principles underlying the work or skill exception.

2 Constitutional Protection

In contrast to English law, US courts have recognised property rights to protect dead bodies and biological materials removed from them. The creation of property rights in US jurisprudence is not, however, dependent on the application of work or skill to transform the separated biological materials. Instead, the corpus of American authority has developed out of the unauthorised removal of organs and the potential protection afforded by the Due Process Clause contained in the Fourteenth Amendment of the US Constitution.

The Fourteenth Amendment protects an individual's right to property against deprivation by the state without due process. Any person alleging a deprivation by the state of any property right without due process possesses a civil cause of action under section 1983 of the Civil Rights Act.[115] To assert a section 1983 claim, a party must establish: (1) that the alleged deprivation was committed by a person acting under colour of state law; and (2) that the deprivation infringed a right, privilege or immunity guaranteed by the US Constitution.[116] The question of

[110] *Kelly*, above n 55, at 631.
[111] *Re Organ Retention*, above n 69, at [148].
[112] Tettenborn, above n 91, at 1024; Grubb, above n 61, at 312; Nuffield Council on Bioethics, *Human Tissue: Ethical and Legal Issues* (Nuffield Council on Bioethics London 1995) [9.11].
[113] See nn 23–122 of ch 5 and accompanying text.
[114] B Nicholas, *An Introduction to Roman Law* (Oxford, Clarendon Press, 1962) 136.
[115] 42 USC 1983.
[116] *Parrat v Taylor* 451 US 527, 535; 101 S Ct 1908 (1981) 1913.

whether an interest amounts to a 'property right' for the purposes of a section 1983 suit is a matter of state law.[117]

The US Court of Appeals for the Sixth Circuit has held that, under the Due Process Clause, family members do have property rights in separated biological materials. In *Brotherton v Cleveland*,[118] the Sixth Circuit considered the constitutional validity of an Ohio statute that permitted a coroner to remove corneas from a deceased person provided the coroner was not aware of any objections from close family members.[119] During an autopsy procedure the deceased's corneas had been removed against the wishes of his wife. The court reviewed the authorities dealing with analogous issues and noted that a majority of cases ruled that 'quasi-property' rights can exist in dead bodies. In this case, the court held that the 'aggregate of rights' given to a spouse under common law and the Uniform Anatomical Gift Act were sufficiently proprietary for constitutional purposes.[120] These rights included the right to possess the body and to control its disposal.

In *Whaley v County of Tuscola*,[121] the Sixth Circuit Court of Appeals considered the validity of a Michigan statute after the deceased's eyes had been removed without consent. The Sixth Circuit again held that the next-of-kin had a 'constitutionally protected property interest' in the dead body of a relative. In explaining the next-of-kin's interest, the court opined that the existence of a constitutionally-protected property right does 'not rest on the label attached to a right granted by the state but rather on the substance of the right'.[122] In this way, both *Brotherton* and *Whaley* emphasised that the rights of the family members should be viewed as a bundle of rights in a proprietary sense. The Ninth Circuit Court of Appeals[123] and several US District Courts[124] have adopted a similar approach to the Sixth Circuit.

The position adopted by the Sixth and Ninth Circuit Court of Appeals has not been widely embraced by other circuit or state courts. In *Georgia Lions Eye Bank Inc v Lavant*,[125] the Georgia Supreme Court held that a statute permitting corneal removal was consistent with the US Constitution because dead bodies are not constitutionally protected property. The court considered that the common law concept of 'quasi-property' did not have any constitutional dimension.[126] Indeed, there is a substantial body of US state court jurisprudence that disagrees with the approach adopted by the Sixth Circuit.[127] Such judicial reluctance to recognise

117 *Board of Regents of State Colleges v Roth* 408 US 564, 577; 92 S Ct 2701 (1972) 2709.
118 923 F 2d 477 (6th Cir 1991).
119 Ohio Rev Code, § 2108.60.
120 *Brotherton*, above n 118, at 483.
121 58 F 3d 1111 (6th Cir 1995).
122 *Whaley*, ibid at 1114.
123 *Newman v Sathyavaglswaran* 287 F 3d 786 (9th Cir 2002) 797.
124 *Hainey v Parrott* (Oh US DC 2005) 6; *Martin v Kim* (Ind US DC 2005) 7.
125 335 SE 2d 127 (Ga SC 1985).
126 *Georgia Lions Eye Bank*, ibid at 128.
127 *Crocker v Pleasant* 778 So 2d 978 (Fla App 2001) 986–88; *Bauer v North Fulton Medical Center* 527 SE 2d 240 (Geo CA 1999) 244; *Culpepper v Pearl St Bldg* 877 P 2d 877 (Colo SC 1999) 880; *Dampier v Wayne County* 592 NW 2d 809 (Mich App 1999) 818.

property rights can be explained at least in part by the attitude of US courts to the Takings Clause in the Fifth Amendment. This clause provides that private property should not be taken for public use without just compensation. Similar issues arise in an environmental context in relation to tradeable pollution rights under the emissions trading regime.[128] US courts have been reluctant to classify tradeable pollution rights as capable of constitutional protection. Again, this reluctance may be best explained as judicial concern over the consequences of applying the Fourteenth Amendment.[129] The approach in *Brotherton* and *Whaley* does not, therefore, represent the more widely accepted position in current US jurisprudence. In general, US courts have not recognised that property rights are created following the separation of biological materials from a dead body.

3 Tort of Wrongful Interference

No English authority supports a tort of wrongful interference with a dead body. The question of whether the tort should be recognised has been the subject of academic debate. Matthews has suggested that if a person has the right to possession of unburied human remains, then that person may bring an action in trespass for unlawful interference.[130] Others have not endorsed the tort.[131] A potential tort of wrongful interference (based on an underlying property right) was recently pleaded in *Re Organ Retention*. The claimants submitted that the elements of the tort were: first, that they had the right to possess their deceased children's bodies, such a right being associated with their duty to effect disposal of the bodies;[132] and, secondly, that the defendants had interfered with that right by retaining or disposing of biological materials without lawful authority.[133] It is unclear from the pleadings whether the cause of action was in trespass or conversion. Given that counsel for the claimants argued that 'it was probably a species of conversion',[134] conversion is analysed here.

The lack of English authority led the claimants to draw on comparative Canadian and Scottish authority. In *Edmonds v Armstrong Funeral Home*,[135] the Alberta Supreme Court considered the legality of a post-mortem undertaken against the widow's wishes. Harvey CJ held that the plaintiff had the right to control the deceased's body and that any unauthorised interference with that right

[128] JE Holloway and DC Guy, 'The Utility and Validity of TDRs under the Takings Clause' (2002) 11 *Pennsylvania State Environmental Law Review* 45, 57.

[129] *Penn Centre Transport Co v New York City* 438 US 104 (1978) 122.

[130] Matthews, above n 5, at 218. See, also: RN Nwabueze, 'Interference with Dead Bodies and Body Parts: A Separate Cause of Action in Tort?' (2007) 15 *Tort Law Review* 63.

[131] IM Kennedy, 'Further Thoughts on Liability for Non-Observance of the Provisions of the Human Tissue Act 1961' (1976) 16 *Medicine, Science and Law* 49, 50.

[132] Particulars of Claim, *AB v Leeds Teaching Hospital NHS Trust & Cardiff & Vale NHS Trust* (High Court of Justice No HQ 0101462 February 2004) [25].

[133] Particulars of Claim, ibid at [26].

[134] *Re Organ Retention*, above n 69, at [152].

[135] [1931] 1 DLR 676 (SC).

provided a cause of action.[136] The cause of action supporting the relief granted in *Edmonds* is not clear. Academic commentary in Canada has subsequently classified the decision as an example of intentional infliction of nervous shock.[137]

The claimants in *Re Organ Retention* also invoked two Scottish authorities which held that the performance of post-mortem examinations and retention of organs without the requisite authorisation might amount to an actionable wrong at common law: *Pollok v Workman*[138] and *Hughes v Robertson.*[139] Again, the identity of the actionable wrong in these cases is elusive. *Clerk and Lindsell on Torts* cites them as examples of the proposition that infringements of a right to possession of a corpse for the purposes of burial are actionable.[140] Whitty demonstrates that this classification cannot be sustained because the English law on conversion and trespass with goods is not part of Scots law.[141] Two other classifications are possible. It has been suggested that these decisions are cases of negligence for nervous shock.[142] Another view regards these cases as examples of the action for *solatium* for wounded feelings arising from affront (the *actio iniuriarum*).[143] Temporary Justice McAulay QC agreed with this latter classification most recently in *Stevens v Yorkhill NHS Trust and Southern General University Hospitals NHS Trust.*[144] It is clear that these Canadian and Scottish cases do not establish a clear line of authority supporting a tort of wrongful interference. Importantly, they are not premised on the existence of property rights. In *Re Organ Retention*, Gage J noted that the decisions cited were cases of unauthorised post-mortems.[145] On the facts of *Re Organ Retention*, Gage J held that the removal of organs and subsequent possession of them was lawful.

By concluding that the defendants' possession of the removed organs was lawful, Gage J avoided the more difficult question of whether the claimants' right to 'possession' of a corpse could be a sufficient interest to found a claim in conversion. This book considers that the 'right to possession' of a corpse does not constitute a property right (as defined in chapter one) because it does not concern a 'thing'. The 'right to possession' cannot therefore be protected by the tort of conversion. Instead, 'the right to possession' is better characterised as a non-proprietary interest and its scope is considered below.

[136] *Edmonds*, ibid at 680.

[137] GHL Fridman, *The Law of Torts in Canada* (Toronto, Carswell, 2002) 81; AM Linden, *Canadian Tort Law* (Toronto, Butterworths, 5th edn, 1993) 51.

[138] (1900) 2 F 354.

[139] 1913 13 SC 394; 1912 SLT 503 (Inner House).

[140] Tettenborn, above n 91, at 1024.

[141] NR Whitty, 'Rights of Personality, Property Rights and the Human Body in Scots Law' (2005) 9 *Edinburgh Law Review* 194, 212–13.

[142] K Norrie, 'Human Tissue Transplants: Legal Liability of Different Jurisdictions' (1985) 34 *International and Comparative Law Quarterly* 442, 463.

[143] Whitty, above n 141, at 199; TB Smith, *A Short Commentary on the Law of Scotland* (Edinburgh, Green, 1962) 655; TW Price, 'Legal Rights and Duties in Regard to Dead Bodies, Post-Mortems and Dissections' (1951) 68 *South African Law Journal* 403, 406.

[144] Above n 70, at [62]–[63].

[145] *Re Organ Retention*, above n 69, at [155].

Despite this characterisation, potential protection through the tort of conversion has been raised in the US by the Second Circuit US Court of Appeals in *Colavito v New York Organ Donor Network Inc.*[146] Peter Lucia died in a hospital in New York and his widow decided to donate both his kidneys to his friend Peter Colavito (the plaintiff), who was suffering from renal disease. Lucia's left kidney was flown to a hospital in Miami where the plaintiff was waiting for its implantation. The right kidney remained in New York. The plaintiff's surgeon discovered that the left kidney was irreparably damaged and unfit for implantation. An employee of the hospital in Miami then attempted to obtain the right kidney from New York but was informed that that the kidney had already been delivered and implanted into another patient. The plaintiff then commenced proceedings in the US District Court for the Eastern District of New York for fraud, conversion and violations of Articles 43 and 43-A of the New York Public Health Law. The District Court granted summary judgment to the defendants and dismissed the conversion claim because 'it is against public policy to recognise broad property rights in the body of the deceased'.[147] On appeal, the Second Circuit affirmed the dismissal of the fraud claim but reserved judgment on the other grounds. Instead, the Second Circuit certified the following questions to the New York Court of Appeals:

(1) Do the applicable provisions of the New York Public Health Law vest the intended recipient of a directed organ donation with rights that can be vindicated in a private party's lawsuit sounding in the common law tort of conversion or through a private right of action derived from the New York Public Health Law?
(2) Does New York Public Health Law immunize either negligent or grossly negligent misconduct?
(3) If a donee can bring a private action to enforce the rights referred to in question 1, may the plaintiff recover nominal or punitive damages without demonstrating pecuniary loss or other actual injury?[148]

In making this certification, the Second Circuit made several important statements. First, the court observed that 'there is by no means a modern consensus that body parts are excluded from conversion actions at common law'.[149] Secondly, the court considered that previous authority concerning the right to possession of a corpse did not answer the question posed by this case. The court recognised that 'property' was not being used here as a legal fiction to claim for emotional distress but instead property was invoked to claim for the actual loss of a separated organ:

> The control-over-corpse cases restrict the recovery of relatives to emotional distress because that is in fact what they suffered. But a lawsuit based on the loss of a donated organ typically seeks more than compensation for injured feelings. The intended recip-

[146] 438 F 3d 214 (2nd Cir 2006).
[147] *Colavito v New York Organ Donor Network Inc* 356 F Supp 2d 237 (EDNY 2005) 246 (Irizarry J).
[148] *Colavito*, above n 146, at 216–17 (Sack and Cabranes JJ; Jacobs J dissenting).
[149] Ibid, at 224.

> ient of a human organ does not bring suit for control over a dead body and its constituent parts. He or she sues for the loss of the functioning organ.[150]

Thirdly, the Second Circuit considered that it is 'arguable that under the New York Public Health Law, a person or entity may have an enforceable property right in a functioning organ'.[151] Further, the court considered that the conversion claim was not answered by the fact that the New York Public Health Law prohibits the sale and purchase of organs:

> But the fact that the State wishes to prohibit the treatment of functioning human organs as though they were commodities does not necessarily imply that it also intends that no one can acquire a property right in them. It does not follow from a law that forbids the sale of a functioning human kidney, that a third party may with impunity take the organ against the express wishes of a potential donor and potential donee.[152]

The New York Court of Appeals accepted the certification on 23 March 2006.[153] The plaintiff died in June 2006[154] before the appeal was heard and Patricia Colavito continued the proceedings as the plaintiff's personal representative. The New York Court of Appeals dismissed the appeal.[155] Rosenblatt J delivered the reasons of the court stating that it was not necessary to consider the statutory cause of action under the New York Pubic Health Law because under § 4302(4) gifts of a deceased donor are conditional on medical benefit to the intended recipient. Here, although the plaintiff was a specified donee, he was someone who could not benefit from either kidney as they were medically incompatible.[156] In relation to the tort of conversion, Rosenblatt J observed that the common law developed 'out of religious and cultural sensibilities involving grave robbery, desecration of corpses and, later on, on unauthorised autopsies'.[157] As a result, common law protection to ensure the 'decent burial for an undesecrated body' was never reliant on a theory of property rights in a dead body.[158] Thus, Rosenblatt J found that the plaintiff here, as a specified donee of an incompatible kidney, had no common law right to the organ.[159] Unfortunately Rosenblatt J did not explain this conclusion in any meaningful way, but he did intimate that the common law protection of the dead body and parts removed from dead bodies may be open to future development. Rosenblatt J noted that: 'considering, however, that the "no property right" jurisprudence was developed long before the age of transplants and other medical advances, we need not

150 Ibid, at 225.
151 Ibid, at 225.
152 Ibid, at 226.
153 *Colavito v New York Organ Donor Network Inc* 6 NY 3d 820 (NY CA 2006).
154 'News in Brief' *New York Law Journal* (28 August 2006) 1.
155 *Colavito v New York Organ Donor Network Inc* 8 NY 3d 43 (NY CA 2006) (Rosenblatt J; Kaye CJ, Ciparick, Graffeo, Read, Smith and Pigott JJ agreeing).
156 Ibid, at 57.
157 Ibid, at 50.
158 Ibid, at 52.
159 Ibid, at 53.

identify or forecast the circumstances in which someone may conceivably have actionable rights in the body or organ of a deceased person'.[160]

4 Summary

The common law has recognised that property rights can be created in biological materials separated from dead bodies. In England and Australia, property rights are created where sufficient work or skill is performed on separated biological materials. As *Dobson*, *Kelly* and *Re Organ Retention* illustrate, the doctrinal underpinnings and external boundaries of the exception demand further consideration and clarification.

English law has not, as yet, embraced a tort of wrongful interference based on conversion. In contrast, the US Sixth and Ninth Circuit of Court of Appeals have granted constitutional protection to the next-of-kin. The US Second Circuit of Court of Appeals has also questioned whether a common law tort of conversion can protect a functioning organ once separated from a dead body. At present, however, the majority of US federal and state courts have declined to find that property rights can be created in biological materials separated from a dead body. Thus, the more substantial authority in the common law world for the creation of property rights is located in—and confined to—the work or skill exception.

C Non-proprietary Interests

The law has also recognised non-proprietary interests in dead bodies and biological materials removed from them. Most notably, this form of protection exists in the right to possession afforded under English and Australian law, which is considered here. Also, the development in US jurisprudence of a 'quasi-property' right is analysed. At first impression it may seem that these authorities should have informed the discussion about property rights. It will be suggested that actions based both on possessory rights and quasi-property concepts are better classified as protecting non-proprietary interests. The analysis also examines protection in Scots law, which is not based on a right to possession of a corpse but instead through tort liability in *solatium* for the relatives' wounded feelings.

[160] *Colavito v New York Organ Donor Network Inc* 8 NY 3d 43 (NY CA 2006) (Rosenblatt J; Kaye CJ, Ciparick, Graffeo, Read, Smith and Pigott JJ agreeing). Following the decision of the New York Court of Appeals in *Colavito*, the Second Circuit Court of Appeals found that Colavito had no cause of action under either the New York common law of conversion or the New York Public Health Law: *Colavito v New York Organ Donor Network* 486 F 3d 78 (2nd Cir 2007) 81 (Jacobs CJ, Cabranes and Sack JJ).

1 Right to Possession for Burial

English common law recognises duties to bury corpses and concomitant rights of possession over corpses. *Sharp v Lush*[161] was the first English case to consider an executor's rights over an unburied corpse. It held that burial expenses fell within the meaning of 'executorship expenses', because '[i]t is part of his [the executor's] official duty to bury the deceased, so that he is liable to pay the funeral expenses without an order'.[162]

The executor's right to possession of unburied corpses was also considered in the oft-cited case of *Williams v Williams.*[163] There, the deceased (Crookendale) directed the plaintiff (Williams) by a codicil to burn his body, with the executors directed to reimburse the plaintiff for the costs he incurred. On Crookendale's death, the deceased's widow and one of his sons obtained the body and had it buried in unconsecrated ground. The plaintiff fraudulently obtained a licence from the Home Secretary to disinter the body on the ground that it was to be interred in the consecrated ground of another cemetery. The plaintiff then sued the executors to recover the costs of the cremation. In dismissing the plaintiff's claim, Kay J found against the plaintiff and in doing so drew on the 'no property' principle derived from the decision of Erle J in *R v Sharpe* (*Sharpe's Case*).[164] Surprisingly, Kaye J did not distinguish *Sharpe's Case* on its facts; whereas *Williams v Williams* involved an unburied corpse, *Sharpe's Case* concerned a buried corpse. This is important because, if *Sharpe's Case* were distinguished, it would not necessarily follow that the 'no property' principle applies to unburied corpses. Kay J also noted that it was the deceased's executors who had a right to possession of the body (although they had no property in it).[165] In support of this finding he referred to *Williams on Executors,*[166] Blackstone[167] and *R v Fox.*[168] Kay J regarded the executor's right as exclusive and relied on it as a further reason to explain why the plaintiff could not lawfully enforce the deceased's wishes. In this way, Kay J used both the 'no property' principle and the executor's right to possession as the twin grounds to defeat the claim.

The right to possession recognised in *Williams v Williams* flows from the duty to dispose of the deceased's body. That duty is considered to be a public duty. In *R v Stewart,*[169] Lord Denman observed that '[w]e have no doubt, therefore, that the common law casts on someone the duty of carrying to the grave the dead body'.[170] The public duty carries with it an enforceable right to possession of the

161 (1879) 10 Ch D 468 (CA).
162 *Sharp*, ibid at 472.
163 (1882) 20 Ch D 659 (CA).
164 (1856–57) Dears & Bell 160; 169 ER 959.
165 *Williams*, above n 163, at 664.
166 (London, Stevens & Sons, 6th edn, 1867) 906.
167 Blackstone, above n 8, at vol II, 508.
168 (1841) 2 QB 246.
169 (1840) 12 Ad & E 773; 113 ER 1007.
170 *Stewart*, ibid at 778; 1009.

body.[171] An executor's enforceable right to possession of a body was affirmed in *Dobson v North Tyneside Health Authority*.[172] There, Peter Gibson LJ referred to the general principle that 'persons charged by the law with interring the body have a right to the custody and possession of it until it is properly buried'.[173]

Courts have since expanded the right to possession to include a coroner's prior right to possession of a body when required for the purpose of coronial inquiries.[174] English common law, therefore, initially allocates the right to possession of the body for burial to the executor of the will of the estate, and not to the surviving spouse or next-of-kin. The right of possession enables the executor to maintain a civil action prior to burial.

If no will exists, then it is not settled under English law what class(es) of other persons, including the relatives of the deceased, will be under a duty to arrange the disposal of the body, and thereby have a right to possession of it. Under English law, when a person dies without a valid will, that person's property is allocated to the Public Trustee until the court appoints an administrator. Thereafter all property is allocated to the administrator.[175] The administrator derives title from the court appointment and cannot act until the appointment.[176] Before an administrator is appointed, there are four potential categories of other persons who may have a right to possess the deceased's body for the purpose of disposal: the person entitled to letters of administration;[177] relatives of the deceased;[178] occupiers of the premises where the deceased passed away;[179] and the local authority.[180]

The potential conflict between the possessory rights of the occupier of the premises where the deceased passed away and the deceased's relatives was considered most recently in *Re Organ Retention*. Prior to *Re Organ Retention*, common law precedent was limited. Although courts identified a right to possession in some circumstances, its nature, content and scope received little judicial attention. In *Re Organ Retention*, however, questions concerning the nature of the right to possession arose from the claimants' argument that English law should recognise a tort of wrongful interference with a human body. The specific argument for a tort of wrongful interference has already been analysed; the argument is considered here with reference to the right to possession.

According to the claimants in *Re Organ Retention*, the first limb of the tort of wrongful interference required that they establish a right and corresponding duty

[171] *R v Fox* (1841) 2 QB 246, 247; 114 ER 95, 96; *R v Scott* (1842) 2 QB 248, 249; 114 ER 97, 97.
[172] [1997] 1 WLR 596 (CA).
[173] Ibid, at 600.
[174] *R v Bristol Coroner; ex p Kerr* [1974] QB 652, 659.
[175] Administration of Estates Act 1925 UK, s 9.
[176] *Ingall v Moran* [1944] KB 160, 168.
[177] Non-Contentious Probate Rules 1987 (SI 1987/2024) r 22(1)(a): App 2. For example, the first three persons with priority, in order, are the deceased's spouse; children; and mother and father.
[178] *Clerk v London General Omnibus Co Ltd* [1906] 2 KB 648 (CA) 663–64; *R v Vann* (1851) 2 Den 325, 326; 169 ER 523, 524; *Jenkins v Tucker* (1778) 1 Bl H 90, 93; 138 ER 55, 57.
[179] *Stewart*, above n 169.
[180] Public Health (Control of Disease) Act 1984 UK, s 46; National Assistance Act 1948 UK, s 50(1).

to possess the body of their child.[181] The argument addressed two aspects of the right: (1) when does the right to possession start and finish? and (2) what is its substantive content? The claimants argued that the common law duty and right to bury a child arises on the death of the child.[182] Flowing from the duty was a right to possession of the child's body in a state that 'was as anatomically complete as was reasonably practicable in the circumstances'.[183] In support the claimants drew an analogy with possession of a car. They argued that such possession encompasses a right to each individual part of the vehicle. Unauthorised removal of the engine would accordingly constitute conversion.[184]

The defendants conceded that a right to possession existed but argued that it did not constitute a freestanding right. Rather, they submitted that such a right would be dependent on the duty to bury. Importantly, the defendants argued that the right only arises when it is reasonably practicable for parents to perform the duty to bury, taking into account: '(i) their means (ii) the state of the body and (iii) their knowledge of the location of the body and any tissue or organs separated from the body'.[185] On this approach, the right did not arise immediately at the time of death because without knowledge, there could be no intention to possess and such intention is necessary to establish a right to possession. As to the substantive content of the right, the defendants claimed that the duty to bury only extended to the body itself and not to any parts of the body lawfully removed from it.[186] To this end, the parents only had a right to bury the 'recognizable body'.[187] On the basis of these propositions the defendants submitted that there had been no interference with the claimants' right to possession for the purposes of burial.

Gage J adopted the defendants' approach in relation to the duty and the corresponding right to bury,[188] accepting that the common law duty was limited and should be construed in the light of the common law 'no property' principle. Following Gage J's approach, the rights that parents can claim at common law extend only to those rights necessary to ensure no breach of the duty to bury. The duty does not arise immediately and, if a person dies in hospital, 'the hospital has the legal right to possess the body at least initially'.[189] At common law, parents lose any right to possession for burial once the body has been buried.[190] As to the content of the duty, Gage J held that it is not necessary that the body be buried in a complete state.[191] Thus, the scope of the right to possession at common law is narrow. It is not equivalent to property rights existing over normal chattels.

[181] *Re Organ Retention*, above n 69, at [128].
[182] Closing Submissions (Claimants) *AB v Leeds Teaching Hospital NHS Trust & Cardiff & Vale NHS Trust* (High Court No HQ 0101462 12 February 2004) (copy on file with author) [9].
[183] Ibid, at [27].
[184] Ibid, at [32].
[185] Closing Submissions (Defendants), above n 78, at [183].
[186] Ibid, at [183].
[187] Ibid, at [208].
[188] *Re Organ Retention*, above n 69, at [135].
[189] Ibid, at [139].
[190] Ibid, at [296].
[191] Ibid, at [158].

Accordingly, the analogy the claimants drew with the possession of a car was flawed. Gage J's analysis is correct because the car (as a chattel at common law) is recognised as a 'thing' in property law, unlike a dead body. The car is therefore capable of being the object of property rights.

The foregoing analysis of the case law is important because it demonstrates that English law protects the right to possession of a body, at least for the purposes of burial. Significantly though, common law protection is not extended any further because the duty to bury is not linked to property rights. The 'right to possession' over a dead body is not a property right analogous to rights over normal chattels. For the law of conversion to apply the object in question must be a thing that is capable of being the subject of property rights. A corpse is not a thing capable of being the subject of property rights at common law. Consequently, it is not capable of being protected through the tort of conversion. The 'right to possession' is, instead, limited to the specific purpose of lawfully disposing of the dead body.

Accordingly, the axiom of 'rights to possession' that courts employ is misleading. Such phraseology should be reserved for property rights. An analogy may be drawn with parents who seek the return of abducted children. At law, parents do not rely on property rights to children for their recovery.[192] Instead, the primary considerations for a court are the right of custody and the best interests of the child.[193] The right recognised by English law over dead bodies is better described as a right to 'custody' of the body for the purposes of burial.

If the right to possession for burial is not proprietary, then it is unclear what interest the common law protects. The common law still seeks to ensure that dead bodies are protected from unwanted interference by third parties, and therefore confers on executors and the next-of-kin rights of control over dead bodies. In the absence of clear authority, there are two possible explanations for the underlying interest protected by the law. The first (and the more openly acknowledged view) is that the duty to bury arises from considerations of public health and safety. On this view it is in the interests of society to ensure that dead bodies are buried and, accordingly, that someone is allocated a right to control a dead body to ensure that the task is performed.

A second (and equally plausible) explanation for the duty and right to bury stems from the respect for human dignity that the law accords to individuals. Respect for the dead has had longstanding importance in how the law addresses the disposal of bodies.[194] The prohibition on disturbance of cadavers (except in exceptional circumstances) reflects the respect that the law provides for individuals. The law recognises that the dignity of individuals is important even after their death; further, the law entrusts the duty to those who are considered most intimately connected with the deceased including the executor, administrator and next-of-kin. These persons are most likely to afford the deceased due respect

[192] C Barton and G Douglas, *Law and Parenthood* (London, Butterworths, 1995) 22.
[193] *Re V-B (Abduction: Custody Right)* [1999] 2 FLR 192 (CA).
[194] *Re Blagdon Cemetery* [2002] 3 WLR 603 (Arches Ct of Cant) 608.

through the provision of a proper funeral. At the very least, it is likely that the interests underlying the right to possession of a cadaver in English law are a combination of respect for the human dignity of the deceased, as well as the more instrumental public concerns of health and safety.

2 Quasi-property Rights

The legal protection of cadavers in the US was initially secured through recognition at common law that individuals had a 'quasi-property' right to the possession of a deceased's body. In *Pierce v Proprietors of Swan Point Cemetery*,[195] the deceased was buried in a plot that he had bought during his lifetime. At his death, the plot passed to his daughter. The deceased's widow bought herself a plot in the same cemetery and disinterred her husband's body and reburied it without the daughter's consent. The daughter sought an order to restore the deceased to the original plot.

The Rhode Island Supreme Court ordered the restoration of the deceased's remains to the original plot. Potter J observed that there were no property rights to a dead body. He did consider, however, that a form of 'quasi-property' arose from the duty and right to protect a human corpse from violation.[196] US courts consider that quasi-property rights are allocated, first, to the surviving spouse,[197] and then to the next of-kin.[198] Again, the underlying basis for protecting these rights is not clear. Some US courts have suggested that the quasi-property rights stem from a public duty of decent disposal and other courts have suggested those rights stem from an interest in the body.[199]

Although the basis for the protection of such rights is unclear, US courts have emphasised that the quasi-property notion should not be considered akin to commercial ownership. Rather, quasi-property rights have been couched in more general language. In *Larson v Chase*, the court observed that the fact that an individual has exclusive control over a body for burial leads 'necessarily to the conclusion that it is his property in the broadest and most general sense of that term'.[200] Such broad terminology does not assist in defining the interest underlying the rights or the scope of the rights that the courts are seeking to protect.

Some guidance as to the content of quasi-property rights was provided in *Whitehair v Highland Memory Gardens*.[201] In that case, an individual brought an action seeking damages for the mishandling and loss of several bodies during a process of relocating a cemetery due to highway construction. Miller J outlined the content of the claimant's rights:

[195] 14 Am Rep 465 (RI SC 1881).
[196] *Pierce*, ibid at 676–77.
[197] *Larson v Chase* 50 NW 238 (Minn SC 1891).
[198] *Enos*, above n 31, at 172.
[199] *Snyder v Holy Cross Hospital* 352 A 2d 334 (Md App 1976) 341.
[200] *Larson*, above n 197, at 239.
[201] 327 SE 2d 438 (WVa SC 1985).

> The quasi-rights of the survivors include the right to custody of the body; to receive it in the condition in which it was left, without mutilation; to have the body treated with decent respect, without outrage or indignity thereto; and to bury or otherwise dispose of the body without interference.[202]

Miller J's extrapolation suggests that the rights arise out of the need to respect the dignity of the deceased. Despite the emergence of quasi-property rights in the context of cadavers, 'quasi-property' is not a widely-recognised legal classification in US common law. The US Supreme Court has only used the term 'quasi-property' in one case, *International News Service v Associated Press.*[203] There, the Supreme Court held that information constituting news must be regarded as quasi-property.[204] There may not, therefore, be any great significance in labelling an interest as quasi-property.

Moreover, some US courts have questioned such 'quasi-property' protection. In *Culpepper v Pearl Street Building Inc,*[205] the Colorado Supreme Court reasoned that a quasi-property theory does not protect a true ownership interest because it redresses emotional harm rather than physical injury. The court held that such cases are not about the injury to the body itself, but whether the improper actions caused emotional or physical pain or suffering to the surviving family members.[206] This reasoning has been followed in other US jurisdictions, and some US courts are shifting the basis of protection of corpses from 'quasi-property' to the law of tort (considered below).[207]

This analysis highlights significant inconsistencies over the protection of cadavers in different US jurisdictions. In part, these inconsistencies are explicable because the courts are uncertain as to what interest is being protected: is it proprietary, based on the dignity of the deceased, or linked to the emotional harm suffered by close relatives? Indeed, the use of 'quasi' is a fiction. 'Quasi' describes something in terms of something else, which it is not. Some US courts have recognised the fiction.[208] Other US courts have not recognised property rights in biological materials removed from cadavers but have, nevertheless, sought to invoke the protection and the effects that a property classification provides.[209] The use of the epithet 'quasi' demonstrates that the law is protecting something that is not property. Rather, the courts are using the concept of property as a means to protect the dead body.

[202] *Whitehair*, ibid at 441.

[203] 248 US 215; 39 S Ct 68 (1918).

[204] *International News Service*, ibid at 373.

[205] 877 P 2d 877 (Colo SC 1994).

[206] *Culpepper*, ibid at 880.

[207] See, for example: *Biro v Hartman Funeral Home* 669 NE 2d 65 (Ohio App 1995) 72; *Brown v Mathews Mortuary* 801 P2d 371 (Idaho SC 1990) 374; *Carney v Knollwood Cemetery Association* 514 NE 2d 430 (Ohio App 1986) 434–35; *Scarpaci v Milwaukee County* 292 NW 2d 816 (Wisc SC 1980) 820–21.

[208] *Carney*, ibid at 434; *Scarpaci*, ibid at 820–22; *Pettigrew v Pettigrew* 56 A 878 (Pa SC 1904) 879.

[209] *Enos*, above n 31, at 172; *Whitehair*, above n 201, at 441; *Meek v State* 185 NE 899 (Ind SC 1933) 901.

The overview of US authority also reveals that such cases are primarily concerned with situations where the law considered that the wrong person was exercising control over separated biological materials from cadavers. The courts have instead sought to ensure that such control is exercised by a more appropriate person. The underlying interest that the courts are protecting in these cases is not proprietary; rather, it is the maintenance of the dignity of the deceased and protection of family members from the infliction of emotional distress.[210] This book has already suggested that these non-proprietary interests may be more appropriately labelled a right to 'custody' of the body for the purposes of burial.

3 Statutory Rights to Possession

In addition to the common law, non-proprietary interests also exist under statute. The discussion here is not intended to provide a detailed overview of these relevant statutory regimes. Instead, the analysis simply demonstrates that certain statutes have created limited rights to possession of cadavers. The HTA 2004 is examined in chapter four.

The first example in English law of a statutory right to possession is the coroner's right to retain the body for a lawful inquest. Under the Coroners Act 1988 UK the coroner has a statutory duty to hold an inquest where there is reasonable cause to suspect that the deceased died from a violent or unnatural death, or died a sudden death of which the cause remains unknown.[211] The coroner's right to retain the body and its parts arises once the decision is made to hold the inquest and ceases at the conclusion of the inquest.[212] Under Australian legislation, a coroner may remove and retain tissues and organs at autopsy without requiring the explicit consent of the family.[213] The common law right to possession of a dead body is therefore subject to the coroner's statutory right.

The English Coroners Act is silent on what control the coroner retains over separated biological materials once the duty to hold an inquest has been discharged. The Coroners Rules were amended in February 2005 so as to regulate more specifically the provision of material separated from bodies.[214] Rules 9, 9A, 12 and 12A provide that where a pathologist or other person retains material following a post-mortem or special examination, that person is required to inform the coroner that material has been retained and the coroner must specify the period for

[210] See, particularly: *Carney*, above n 207, at 434–35; *Pierce*, above n 195, at 681; *Reinihan v Wright* 25 NE 822 (Ind SC 1890) 824–25.

[211] Coroners Act 1988 UK, s 8(1).

[212] *R v Bristol Coroner*, above n 174.

[213] Human Tissue Act 1983 (NSW) ss 23, 24, 25; Human Tissue Act 1985 (Tas) ss 23, 24, 25; Human Tissue Act 1982 (Vic) ss 25, 26, 27; Human Tissue and Transplant Act 1982 (WA) ss 25, 26, 27; Human Tissue Transplant Act 1979 (NT) ss 18, 19, 20. For a discussion of the legal issues in relation to the conduct of autopsies and the work undertaken by coroners, see: I Freckelton and D Ranson, *Death Investigation and the Coroner's Inquest* (Oxford, OUP, 2006) 353–85.

[214] Coroners (Amendment) Rules 2005 (No 420, 28 February 2005). The amendments came into force on 1 June 2005: Coroners (Amendment) Rules 2005, r 1(1).

which the material must be preserved.[215] The period for which material can be preserved cannot extend beyond the date on which the coroner's functions cease.[216] The coroner must also notify certain persons[217] of the period for which the material is to be preserved and the options for dealing with the material when that period expires. The options include: (1) disposal of the material by burial, cremation or other lawful disposal by the pathologist; (2) return of the material to the relevant person on request; and (3) retention of the material with the consent of the relevant person for medical research or other purposes.[218] Despite providing for notification procedures, the Coroners Rules do not provide sanctions in the event that the procedures are not complied with. Further, neither the Coroners Act nor the Coroners Rules classify removed materials as property. Instead, their legal status, and any rights attaching to them, remains undefined.

A second example of a statutory right to possession was located in the HTA 1961 (repealed by the HTA 2004). The HTA 1961 was intended to remove any doubt about the legitimacy of using removed biological materials for medical education and research.[219] Under (the now-repealed) section 1 of the HTA 1961, where certain conditions were met, the 'person lawfully in possession' of the body of a deceased person could authorise the removal of its parts for therapeutic purposes, or for the purposes of medical education or research. The HTA 1961 extended to 'parts of the body', including biological materials. In most cases, the 'person lawfully in possession' was the hospital authority.

The HTA 1961 did not clarify whether 'lawfully in possession' had the same meaning as 'right to possession'.[220] Despite the ambiguity of 'lawfully in possession', Dworkin and Kennedy noted that the HTA 1961 provided 'little support for the existence of property rights in the cadaver, or parts of it: references are to persons "lawfully in possession" of the body'.[221] However, once biological materials were removed, the statute did not confer legal protection for that possession. The HTA 1961 therefore created limited rights to possession and control in respect of cadavers and parts removed from them. It did not create property rights to biological materials on separation.

[215] Rules 9 and 9A relate to post-mortems and rules 12 and 12A concern special examinations.

[216] Coroners Rules 1984, rr 9(9), 9A(4), 12(8) and 12A(4).

[217] Under rr 9(6), 9A(3), 12(5), 12A(3) of the Coroners Rules 1984 the coroner must notify: (a) one of the persons referred to in r 20(2)(a) (defined as 'a parent, child, spouse, partner and any personal representative of the deceased'); and (b) any other relative of the deceased who has notified the coroner of his desire to attend, or to be represented at the post-mortem examination.

[218] Coroners Rules 1984, rr 9(9), 9A(4), 12(8) and 12A(4).

[219] *Hansard* HC vol 632 col 1232 (20 December 1960).

[220] During the second reading debate, the Minister for Health observed that defining who was 'lawfully in possession' was a difficult matter because 'in law there is no property in a dead body': *Hansard* HC vol 632 col 1254 (20 December 1960).

[221] G Dworkin and I Kennedy, 'Human Tissue: Rights in the Body and its Parts' [1993] 1 *Medical Law Review* 291, 298.

4 Protection under Tort Law

Throughout the twentieth century, some US state courts and Canadian courts moved away from the notion of 'quasi-property' rights to a dead body, instead favouring the award of damages to the next-of-kin who had suffered emotional distress from the improper treatment of their relative's corpse.[222] One commentator has suggested that US common law now recognises five distinct causes of action in tort in this area: (1) intentional infliction of emotional distress; (2) intentional mishandling of a dead body; (3) abuse of a dead body; (4) negligent infliction of emotional or mental distress; and (5) negligent or wrongful interference with a dead body.[223] The first three causes of action are similar and require the plaintiff to show outrageous, wilful or wanton conduct by the defendant. These requirements are not easily established.

The American Law Institute's *Restatement of the Law, Second, Torts* purports to condense tort and quasi-property forms of action into a single cause of action in tort. It provides:

> [O]ne who intentionally, recklessly or negligently removes, withholds, mutilates or operates upon the body of a dead person or who prevents its proper interment or cremation is subject to liability to a member of the family who is entitled to disposition of the body.[224]

The commentary to the Restatement notes that the 'technical basis' of the cause of action is 'interference with the exclusive right of control of the body'.[225] Even though this right has been called a 'property' or 'quasi-property' right by certain courts, the Restatement notes that these rights do not easily fit within the law of property.[226] A tortious cause of action is based on a survivor's personal rights, rather than proprietary rights. Under tort theory, damages reflect the emotional harm suffered by relatives instead of compensation for the non-consensual transfer of property rights. In *Scarpaci v Milwaukee County*,[227] for example, the Wisconsin Supreme Court observed that Wisconsin law clearly provides that the family of the deceased has a legally recognised right to entomb the deceased's remains with respect for his integrity and without mutilation.[228] The court held that the basis for recovery of damages should not be found in the property rights to a dead body, but in the personal right of the family of the deceased to bury the

[222] *Carney*, above n 207.

[223] RN Nwabueze, 'Biotechnology and the New Property Regime in Human Bodies and Body Parts' (2002) 24 *Loyola of Los Angeles International and Comparative Law Review* 19, 29.

[224] American Law Institute, *Restatement of the Law, Second, Torts* (Washington, American Law Institute, 1965) § 868; US Congress Office of Technology Assessment, *New Developments in Biotechnology: Ownership of Human Tissue and Cells—Special Report* (Washington, US Government Printing Office, 1987) 73.

[225] *Restatement of the Law, Second, Torts*, ibid at [275], [868].

[226] Ibid.

[227] 292 NW 2d 816 (Wisc SC 1980).

[228] *Scarpaci*, ibid at 820.

body.[229] In *Mason v Westside Cemeteries Ltd*,[230] a Canadian court in Ontario found that the plaintiff was entitled to damages for mental distress for negligence after a cemetery had misplaced the ashes of the plaintiff's father.[231] Following this line of authority, the law should be concerned primarily with any improper acts and whether such acts caused emotional or physical harm to the surviving relatives.

Section 868 of the Restatement is not a statutory provision binding US courts. As a result, US courts still require plaintiffs to prove the requirements under the separate causes of action identified above. Courts have, for example, declined to follow section 868 in the context of organ donation.[232] Further, causes of action in tort are not without their difficulties. The major obstacle faced by plaintiffs is that, in most cases, courts have regularly held that plaintiffs must have suffered physical harm or physical loss. This obstacle explains why US plaintiffs have often sought alternative recourse through the doctrine of quasi-property.[233] The quasi-property doctrine does not require the plaintiff to show wilful or wanton misconduct by the defendant or to establish physical or pecuniary loss.[234]

These developments in US and Canadian tort law are nonetheless significant, and may provide a basis for the future protection of non-proprietary interests in separated biological materials under English law. Part II returns to the potential development of a tort analogous to the American tort of intentional infliction of emotional distress.

5 Solatium for Affront

In Scotland, the availability of compensation for *solatium* for wounded feelings (*actio iniuiarum*) warrants consideration. This action was considered most recently in *Stevens v Yorkhill NHS Trust and Southern General University Hospitals NHS Trust*,[235] where a pursuer sought £100,000 damages following the non-consensual removal and retention of the brain of her baby at post-mortem. Her daughter was born on 30 June 1995, but died on 1 August 1995 following multiple organ failure. A post-mortem was conducted but the pursuer was not informed that her daughter's brain would be removed and retained. It was not until October 2000 that the pursuer received a letter from Yorkhill NHS Trust advising that her daughter's brain had been removed and retained in the course of a post-mortem. The pursuer claimed that she had suffered psychiatric injury as a consequence and at trial relied on two causes of action: negligence and the action of *solatium* for affront.

229 *Scarpaci*, ibid at 820.
230 (1996) 135 DLR (4th) 361.
231 *Mason*, ibid at 380.
232 *Ramirez v Health Partners of S Ariz* 972 P2d 658 (Ariz App 1998) 665–67.
233 *Carney*, above n 207, at 434.
234 R Rao, 'Property, Privacy and the Human Body' (2000) 80 *Boston University Law Review* 359, 385–86.
235 Above n 70.

First, Temporary Judge CJ MacAulay QC found that, even though there was no doctor-patient relationship at the time consent to the post-mortem was sought, a duty of care was owed to the pursuer to disclose the fact that organs may be removed and retained during the post-mortem.[236] Secondly, Temporary Judge MacAulay found that Scots law recognised the action of *solatium* for affront for the unauthorised removal and retention of organs from a dead body.[237] In reaching this conclusion, MacAulay J drew on three Inner House cases—*Pollok v Workman*,[238] *Conway v Dalziel*[239] and *Hughes v Robertson*[240]—as examples of liability for *solatium* for the relatives' wounded feelings arising from the unauthorised and wrongful removal and retention of organs at post-mortem.[241] Whitty had argued previously for such a construction of these authorities.[242]

In *Pollok*, for example, a doctor performed a post-mortem without the consent of the relatives or a legal warrant. The pursuer was one of the deceased's daughters and she had arranged for the body to be placed in a house before the funeral. The daughter instructed her solicitor to commence proceedings against the deceased's employer for compensation and against the employer's solicitors (and their insurers) who had instructed the doctor to perform a post-mortem. The daughter sued the doctor, solicitor and insurers in *solatium* for hurt feelings following the wrongful interference with the corpse of her father. Lord Kyllachy held that the daughter had title to sue because 'the cutting up or dissecting, or other unauthorised mutilation of a near relative's body, constitutes a wrong of which a near relative has a title to complain'.[243] Temporary Judge MacAulay observed that '[i]t was that affront that Lord Kyllachy concluded justified a claim for *solatium* for "injured feelings" '.[244] Although the Second Division found that the action was competent, it failed because other possible claimants had not been joined. In *Conway v Dalziel*,[245] the Lord Ordinary approved the decision in *Pollok* but declined to follow it on technical grounds. On appeal, Lord Adam noted that the removal of organs during a post-mortem 'is a much more serious wrong than the post-mortem examination itself'.[246] Temporary Judge MacAulay noted that *Conway*:

> [P]rovides clear authority for the proposition that the unauthorised removal and retention of organs from the body of a deceased person constitutes an independent legal wrong in respect of which relatives can sue for *solatium*. Again, it is the notion of affront caused that seems to be the basis of the claim.[247]

[236] *Stevens*, ibid at [87].
[237] Ibid, at [62].
[238] (1900) 2 F 354.
[239] (1901) 3 F 918.
[240] 1913 SC 394; 1912 SLT 503 (Inner House).
[241] *Stevens*, above n 70, at [63].
[242] Whitty, above n 141, at 216.
[243] *Pollok*, above n 138, at 355.
[244] *Stevens*, above n 70, at [34].
[245] (1901) 3 F 918.
[246] *Conway*, ibid at 919.
[247] *Stevens*, above n 70, at [39].

Finally, in *Hughes v Robertson*[248] the wife and children of a deceased sued a doctor for performing a post-mortem examination on the body of the deceased without their consent. The pursuers alleged that during the course of the post-mortem biological materials were removed and destroyed. Lord MacKenzie indicated that, although the case was not appropriate for a jury trial, it raised the same wrong as had been committed in *Pollok*. Temporary Judge MacAulay noted that *Hughes* therefore 'lends some support to the line taken in *Conway v Dalziel* that the removal and retention of organs can itself constitute a separate and independent legal wrong'.[249]

On the basis of this analysis of the case law, Temporary Judge MacAulay concluded that there is a Scots common law rule that the unauthorised removal or retention of organs at post-mortem constitutes an independent legal wrong in respect of which relatives can sue for *solatium* for affront.[250] Importantly, he also noted that this doctrine potentially provided a lower threshold for recovery than found in the tort of negligence:

> In principle solatium for 'hurt feelings' caused by affront based upon the actio injuriarium is a different animal to the solatium that can be awarded to a claimant for physical or psychiatric injury. Prima facie the threshold for recovery for hurt feelings is lower than that for psychiatric injury. Both parties seemed to proceed on the basis that solatium is the same whether the claim is based upon an independent wrong or negligence, but I am not convinced that that is correct. Accordingly, and in the absence of detailed submissions on this issue, I would not have refused probation on this head of claim even if the pursuer fell foul of the tests necessary to sustain a claim for psychiatric injury based on negligence.[251]

Stevens provides an interesting comparison with the common law authority that confers a right to possession for the purposes of burial. Temporary Judge McAulay noted that 'English law, with its different legal history may not recognize the existence of such a wrong, but that does not impact upon the position in Scotland'.[252] English law, through the right of possession, seeks to protect the dignity of the deceased. Arguably, the Scots law cause of action protects the dignity of the relatives. In the earlier Scottish cases it was not clear whose dignity, that of the deceased or that of the bereaved relatives, was impaired by unauthorised post-mortems.[253] In Roman-Dutch law the right of the deceased's relatives to *solatium* for interference with the deceased's body appears to depend on the notion of an indirect *iniuria* (being insult or affront).[254] An *iniuria* directed to A becomes an

[248] 1913 SC 394, 399.
[249] *Stevens*, above n 70, at [43].
[250] Ibid, at [62].
[251] Ibid, at [63].
[252] Ibid, at [62].
[253] Whitty, above n 141, at 218–19.
[254] M Earle and NR Whitty, 'Medical Law' (2006) *Laws of Scotland, Stair Memorial Encyclopedia Reissue*, [355].

iniuria against B because of the relationship of B to A.[255] On this approach it was not, therefore, an element of a cause of action that the pursuer suffer wounded feelings. Whitty has observed that on normal principles of delict, 'it is thought that if there are no wounded feelings, there is no right of action for iniuria'.[256] Indeed, Temporary Judge MacAulay made it clear in *Stevens* that he considered that the dignity invaded was that of the deceased's relatives:

> [W]hat appears to lie at the heart of those decisions is the fact that near relatives suffered affront by the unauthorised actings. The Judges in the Scottish cases considered that, in the circumstances, the unauthorised post-mortems and the unauthorised removal and retention of body parts, disclosed such an insensitivity to the feelings of near relatives following upon the death of a loved one, that such conduct constituted an affront to their dignity as relatives of the deceased so as to justify being classified as a civil wrong in which damages by way of solatium could be claimed.[257]

As with the American tort law cases, the focus lies on the emotional distress suffered by the relatives of the deceased.

6 Respect for Private and Family Life

A separate cause of action potentially exists under the HRA, which incorporates the substantive rights provisions of the ECHR into English law.[258] Article 8(1) of the ECHR provides, *inter alia*, that each individual has a right to respect for private and family life. It is well established in ECtHR jurisprudence that the concept of 'private life' includes a person's physical and moral integrity.[259]

In English law, the protection potentially afforded by article 8 was addressed in *Re Organ Retention* in relation to a hypothetical factual matrix. The parties hoped that judicial guidance on the three issues of law raised by the factual matrix would assist in resolving the other claims entered on the NOGL group litigation register, aside from the three lead cases. Gage J set out the factual matrix in his judgment:

> The Factual Matrix
>
> (i) A pathologist lawfully acquires in the context of a coroner's post-mortem and thereafter possesses a brain taken from a child. Neither the removal nor retention of the brain is known to the child's parents.
>
> (ii) The child is buried before the fixing of the brain is completed.

[255] C Amerasinghe, *Aspects of the Actio Iniuriarum in Roman-Dutch Law* (Colombo, Lake House, 1966) 203; J Neethling, JM Potgieter and PJ Visser, *Neethling's Law of Personality* (Durban, Butterworths, 2nd edn, 2005) 67–69.

[256] Whitty, above n 141, at 219.

[257] *Stevens*, above n 70, at [57].

[258] Lord Hope observed in *Ex parte Kebilene* [2000] 2 AC 326 (HL) at 374 that '[i]t is now plain that the incorporation of the European Convention on Human Rights into our domestic law will subject the entire legal system to a fundamental process of review and, where necessary, reform by the judiciary'.

[259] *X and Y v Netherlands* [1985] ECHR 4, [22].

(iii) The coronial purpose for which the brain was acquired and fixed then comes to an end.
(iv) Thereafter, and without recourse to the child's parents, the pathologist decides to use and uses the brain for a research project.

The Assumptions

(i) The decision to fix and/or the fixing of the child's brain prior to its use in the research project constitutes sufficient application of work or skill so as to render the brain capable of ownership.
(ii) The research project decided upon and undertaken by the pathologist is solely in the public interest.[260]

The parties asked the court to consider whether the decision to use the child's brain in the research project constituted an infringement of the right to respect for private and family life conferred by article 8 of the ECHR. Gage J held that article 8(1) had been infringed and further, he considered that the research was unlikely to be capable of justification under article 8(2), even though he noted that each case will turn on its own facts.[261] Although such research may be in the public interest, it is difficult to argue that it is necessary for the protection of public health (a specific exception listed in article 8(2)).

7 Summary

Developments in the common law now protect non-proprietary interests over biological materials removed from dead bodies. Examples in English and US authorities illustrate how courts have employed terms such as 'right to possession' and 'quasi-property' in an attempt to describe the rights that arise from the duty to bury. These rights do not protect property rights. Looking beneath the surface, these terms are instead used as an instrumental means to provide control over a human body. Conferring control is intended (at least in part) to ensure that the dignity of the deceased is respected, and the law appears to consider that this control is best allocated to those who are most closely connected to the deceased—the executor, next-of-kin and other family members. It is more appropriate, therefore, to refer to the right as a right to 'custody' of the dead body for the purposes of burial.

D Conclusion

This chapter has focused on three key developments in the law. First, the starting point for the common law in England, Australia and the US is the 'no property'

[260] *Re Organ Retention*, above n 69, at [288]–[89].
[261] Ibid, at [298].

principle. The second development in English and Australian law is the creation of property rights under the work or skill exception. Fundamental questions about the nature and scope of the exception remain unsettled.

The third development concerns the legal protection of non-proprietary interests. English, Australian and US jurisdictions all recognise the right of possession and associated duty of either the executor or next-of-kin to bury cadavers. Both rights and duties derive from a respect for human dignity and public notions of religion, health and decency. The chapter also examined non-proprietary protection through the law of tort. In England, a potential tort of wrongful interference was raised in *Re Organ Retention*, and awaits further judicial exploration. In the US, by contrast, there is a more developed body of jurisprudence. To conclude, the chapter examined Scots law and liability in *solatium* for the relatives' wounded feelings. This action does not stem from a right to possession, but the relatives' right to dignity.

It remains to be considered whether physical transformation of biological materials by work or skill is the only circumstance that can create property rights. Importantly, common law decisions applying the work or skill exception have been careful to leave open the question of whether biological materials separated from a living person must be transformed in some way for the law of property to apply. In *Kelly*, Rose LJ specifically indicated that the application of work or skill may not be necessary to create property rights. The next chapter assesses how the common law and statutory regimes in England, Australia and the US have addressed the increasingly significant question of what property rights apply to bodily materials removed from living persons.

3

Living Body

Common law and civil law jurisdictions deal differently with the creation and allocation of property rights in relation to bodily materials separated from a living person. German law provides, for example, that property rights to bodily materials are allocated to the living person from whom materials are removed.[1] Section 90 of the Bürgerliches Gesetzbuch (BGB) (the German Civil Code) sets out that only tangible objects can be 'things' and, therefore, the subject of property rights. The German Federal Court of Justice has accepted that separated bodily materials are 'things' for the purposes of the BGB.[2] The picture is different in the common law world. As Laurie observes: 'while property rights are routinely granted in respect of human material, this is usually done to the exclusion of the one person who is central to the entire enterprise, namely, the individual from whom the material is taken'.[3]

This chapter commences by considering the legal status of slaves in order to understand the historical underpinnings of the common law. Next, it critiques the US common law authority that has explicitly rejected an individual's property rights to separated biological materials. Although some statutes and common law decisions appear to recognise property rights to separated bodily materials, they do not represent a coherent body of jurisprudence. Thirdly, the chapter identifies principles that have been developed by legislatures and the courts to recognise property rights in this context. The chapter concludes by exploring non-proprietary interests.

[1] Bürgerliches Gesetzbuch (BGB) (German Civil Code) § 90 and Strafgesetzbuch (StGB) (German Criminal Code) § 242. See, also: J Taupitz, 'The Use of Human Bodily Substances and Personal Data for Research: The German National Ethics Council's Opinion' (2006) 3 *Journal of International Biotechnology Law* 25.

[2] It is accepted by commentaries on the BGB, as well as the German Federal Court of Justice, that separated bodily materials are 'things' for the purposes of the BGB: HC Jaurnig (ed), *Kommentar zum StgB/BGB* (Verlag Beck München 2004) § 90/242, 33 and BGHZ 1994, 124 (German Federal Court of Justice) 53. Further, § 242 of the Strafgesetzbuch (StGB) (the German Criminal Code) provides that theft is committed when a 'thing' belonging to another is taken with the intention of appropriating it, which would include separated bodily materials.

[3] GT Laurie, *Genetic Privacy* (Cambridge, CUP, 2002) 304.

A Slavery

The starting point for any analysis of the common law in respect of living persons lies with the legal treatment of slaves—the first context in which the question of property rights and a living person arose. In Roman law, although 'nobody [was] regarded as owner of his own body-parts',[4] slaves were considered to be chattels and not legal persons.[5] Where slavery was historically lawful, slaves were therefore the object of property rights. As Salmond observed in England: '[l]ike cattle, [slaves] are things and the object of rights; not persons and the subject of them'.[6] There are very few cases addressing slavery in English law. *Gregson v Gilbert*[7] concerned the pushing overboard of slaves from a ship that was running short of water. In that case, Solicitor-General Lee noted that 'a portion of our fellow-creatures may become the subject of property. This, therefore, was a throwing overboard of goods'.[8] The question for the court was whether there was necessity for the act. Lord Mansfield did not adopt the statement by Solicitor-General Lee directly, but he did so indirectly by proceeding to address the necessity defence. The classification of slaves as property disappeared with *Sommersett's Case*,[9] where the court held that the English tradition of *habeas corpus* meant that a slave being kept in England (awaiting sale overseas) should be set free.[10] Subsequent statutory intervention prohibited ownership of living persons as property.[11] Although the early common law considered that slaves could be regarded as chattels, English law thus moved to recognise that a living person should not be considered to be property. As the rest of this chapter demonstrates, this fundamental premise has indelibly influenced the development of the law concerning separated bodily materials.

B Rejection of Individual Property Rights

In the common law world to date, only US courts have been asked to consider directly whether a living person can claim property rights in separated biological materials. English and Australian courts are yet to address this difficult question.

[4] Ulpian, *Edict, book 18*, in the *Digest of the Justinian* A Watson (ed), (Philadelphia, University of Pennsylvania Press, 1985) D.9.2.13.pr. Quoted by Lord Rodger in *R v Bentham* [2005] UKHL 18; [2005] 1 WLR 1057, [14].

[5] Slaves were classified as things and could be bought and sold: Gaius *Institutes* 1.9. A slave was termed *servilis persona* and in Roman law, the slave was the 'subject of a right residing in his master (not against himself, but against third persons)': R Campbell (ed), J Austin, *Austin's Lectures on Jurisprudence or the Philosophy of Positive Law* (London, Murray, 5th edn, 1885) vol 1, 387.

[6] JW Salmond, *Jurisprudence* (London, Stevens & Haynes, 1902) 334.

[7] (1783) 3 Dougl 232; 99 ER 629.

[8] *Gregson*, ibid, at 629–30.

[9] (1772) Loft 1; 98 ER 499.

[10] Ibid, at 19; 510.

[11] This process began in 1807 with the banning of the African slave trade (47 Geo III c 36). It was completed in 1834 (3 & 4 Will IV c 73, s 12).

1 *Moore v Regents of the University of California*

Moore v Regents of the University of California[12] is the most well-known common law decision concerning property rights and human biological materials. Indeed, *Moore* is often taken as the starting point for consideration of the legal status of human biological materials by law reform bodies[13] and academic commentaries in England, Australia and the US.[14] This is because *Moore* was the first (and remains the principal) common law authority to address the question of whether or not a living person can claim property rights to separated biological materials.

The case involved an action brought by John Moore against five defendants: (1) Dr David Golde; (2) the Regents of the University of California; (3) Shirley Quan; (4) Genetics Institute Inc; and (5) Sandoz Pharmaceuticals Corporation and related entities. In 1976, Moore visited the Medical Centre of the University of California (UCLA Medical Centre) after learning that he had leukaemia. His physician, Golde, confirmed the diagnosis and recommended the removal of Moore's spleen. Moore signed a written consent form authorising the splenectomy, but at no stage did Golde seek Moore's consent or inform Moore of his intention to conduct further research on the spleen. Moore's spleen was removed and, over the next seven years, Moore returned to the UCLA Medical Centre to provide samples of blood, blood serum, skin and bone marrow aspirate.

Throughout this period, Golde and Quan conducted research on Moore's T-lymphocyte cells. Moore's T-lymphocytes were valuable because they overproduced certain lymphokines, making the corresponding genetic material easier to identify. One of Golde and Quan's objectives was to establish a cell-line from Moore's T-lymphocytes. Although the researchers had access to Moore's primary cells, the cells usually died after a few days. By 1979, the researchers had developed Moore's primary cells into a cell-line. A great potential economic value lay in the process that they developed to produce the cell-line and, in 1984, the Regents were granted a patent. The patent included methods for using the cell-line to produce lymphokines, and it was subsequently estimated that the cell-line had a commercial value of US$3 billion dollars over a six-year period.[15] The Regents then entered into a series of commercial agreements with Genetic Institute and Sandoz conferring rights to the cell-line and its products. It was not until 1983 that Moore was asked to sign a consent form in respect of the continuing research, which he

[12] 249 Cal Rptr 494 (Cal App 1988) (*Moore* CA) and 793 P 2d 479 (Cal SC 1990) (*Moore* SC).

[13] Nuffield Council on Bioethics, *Human Tissue: Ethical and Legal Issues* (Nuffield Council on Bioethics, London, 1995) [9.12].

[14] B Dickens, 'Donation and Transplantation of Organs and Tissues' in A Grubb (ed), *Principles of Medical Law* (Oxford, OUP, 2nd edn, 2004) 1036 (England); RS Magnusson, 'Proprietary Rights in Human Tissue' in N Palmer and E McKendrick (eds), *Interests in Goods* (London, LLP, 2nd edn, 1998) 50 (Australia); R Gold, 'Owning Our Bodies: An Examination of Property Law and Biotechnology' (1995) 32 *San Diego Law Review* 1167, 1169 (United States).

[15] *Moore* SC, above n 12, at 482. John Moore's cell-line can be ordered through the American Tissue Culture Catalogue at http://www.atcc.org (accessed on 1 June 2007). Moore's cells are for sale as CLR-8066.

refused to do. The consent form was set out in the Court of Appeal's judgment,[16] and although it was primarily concerned with the use of body products, it contained a clause referring to 'rights' in any cell-lines and other products developed, and asked the patient to grant these rights to the Regents.

Moore brought proceedings against the five defendants stating thirteen causes of action, including an action for breach of fiduciary duty and for the tort of conversion (relating to the appropriation of Moore's property rights in his spleen). The conversion claim was brought against all the defendants, and the Superior Court only considered the validity of the conversion cause of action. Demurrers were sustained by the Superior Court and Moore appealed to the Californian Court of Appeal. A majority of the Court of Appeal (Rothman AJ and Woods PJ; George AJ dissenting) held that Moore had property rights in his excised organ sufficient to maintain the cause of action for conversion.[17]

The majority opinion, delivered by Rothman AJ, commenced by defining conversion as 'a distinct act of dominion wrongfully exerted over another's personal property'.[18] On the facts, the majority considered that the plaintiff's spleen was something over which the plaintiff enjoyed unrestricted rights of use, control and disposal. These rights, they held, were proprietary. The majority drew on three sources to support their view. First, they cited *Venner v State of Maryland*,[19] where rights of ownership and control over waste products were considered proprietary.[20] Secondly, they cited cases establishing a limited property interest over dead bodies for the purposes of burial.[21] Thirdly, they noted statutes and cases establishing property rights over the use of a deceased's corneas.[22] These strands of authority, the majority claimed, all rested on the assumption of a general proprietary right: that individuals have the ultimate right of control over their bodies.[23] The majority acknowledged that policy considerations regarding the impact on medical research were the major obstacle to the recognition of an individual's property rights in their own bodies. They concluded that there was no reason to suggest that medical research would suffer by requiring the consent of the donor.[24]

Although the majority considered that the plaintiff had a right of control and hence a proprietary right in his bodily materials, their reasoning is not persuasive for several reasons. They failed to link this right of control with traditional notions of property. They did not explicitly identify what underlying individual interest justified the creation of property rights. At one point, the majority indicated that the individual interest stemmed from the right to privacy. Rothman AJ observed that, if patients could not control what happened to tissues removed from them,

[16] *Moore* CA, above n 12, at 531–32.
[17] Ibid, at 503.
[18] Ibid.
[19] 554 A 2d 483 (Md App 1976).
[20] *Moore* CA, above n 12, at 505.
[21] Ibid, at 506.
[22] Ibid.
[23] Ibid, at 503
[24] Ibid, at 508.

it would open the 'door to a massive invasion of privacy'.[25] Notwithstanding this assertion, the majority did not address in any detail the role of privacy as a justification for property rights.

On appeal, a majority of the Californian Supreme Court (Lucas CJ, Panelli, Eagleson, Kennard, and Arabian JJ; Broussard and Mosk JJ dissenting) reversed the decision of the Court of Appeal. The majority held that, although Moore had a cause of action for Golde's failure to obtain his informed consent, he did not have a cause of action in the tort of conversion. At the outset, the majority noted that no court had previously used the tort of conversion in relation to human cells in medical research.[26] As a result, the majority asked two questions: (1) did the tort of conversion give Moore a cause of action under existing law? and (2) if not, would an extension of the tort of conversion be justified?

For Moore to succeed under existing law, the majority reasoned that he would need to have retained an ownership interest in his cells following their removal. The majority found, however, that Moore had not retained any ownership interest, because laws governing such things as biological materials dealt with these materials as *sui generis*.[27] Furthermore, the majority held that Californian statutory law limited a patient's continuing interest in their cells. To this end, the majority referred to section 7054.4 of the Health and Safety Code, which provided for the destruction of samples following scientific use.[28]

The majority also specifically considered Moore's claim to the patented cell-line and products derived from it. They opined that the cell-line was both factually and legally distinct from the cells taken from Moore's body.[29] In addition, they held that Moore's claim over the cell-line was inconsistent with the patent, which constituted an authoritative determination that the cell-line was the product of invention.

The Californian Supreme Court majority's analysis is unconvincing. Even if a statutory regime strips away some proprietary characteristics, it does not necessarily follow that any remaining right is not of a proprietary character. Legislation often imposes limits and restrictions over proprietary rights, such as in the case of planning laws. Moreover, the claim that the cell-line was legally distinct because it was the subject of a patent is misguided. Moore's claim was directed to the conversion of his biological materials prior to the grant of the patent. As Mosk J observed in dissent, if Moore did have property rights in his cells, then the patenting of them would not alter his property rights.[30] The majority ought to have separated the question of rights as between the biological materials and the cell-line and in doing so clarified the different questions pertaining to each. A cell-line is arguably a new item created from the original biological materials; therefore any rights pertaining to the cell-line are not inhibited

[25] Ibid.
[26] *Moore* SC, above n 12, at 487.
[27] Ibid, at 489.
[28] Ibid, at 491.
[29] Ibid, at 492.
[30] Ibid, at 511.

by a source's underlying property rights or non-proprietary interests to the original biological materials.

Next, the majority addressed the second question as to whether liability in conversion should be extended. The majority gave three reasons as to why an extension of liability would be inappropriate. First, a balancing of policy considerations did not support extending the tort.[31] Secondly, any such extension would be more appropriately achieved by legislative intervention, not judicial activism.[32] Thirdly, the tort was not necessary to protect patients' rights.[33] It is clear from the majority judgment that, of these three reasons, policy considerations were the majority's greatest concern.[34] On the one hand, they recognised the need to protect the right of patients to make autonomous medical decisions, a right which was considered to be grounded in the longstanding principles of fiduciary duty and informed consent. In contrast, the threat of 'civil liability' and the hindering of research weighed heavily against providing a remedy.[35] They noted that scientific research could be hindered both by reducing the volume of biological materials available and by destroying the economic incentive to conduct medical research.[36] The majority held that a balancing exercise was answered by existing disclosure obligations. Those obligations adequately protected patients' privacy rights without hindering research.[37]

Again, the majority's reasoning is problematic. The tenor of the judgment advocates legal protection of the development of medical research because of its social utility, at the expense of prioritising the protection of individual rights. The majority considered individual rights to be adequately protected by the doctrines of fiduciary duty and informed consent. Yet this approach fails to recognise that these doctrines are concerned with the scope and nature of a doctor's duty of care in treating the patient. They are not concerned with the use of biological materials after the treatment is finished. For example, to discharge a duty of care, the concept of informed consent requires the doctor to inform the patient of the nature and extent of risks associated with a particular procedure. The duty does not impose a requirement to inform the patient of the subsequent uses of any separated biological materials.

The majority's other broad concern was economic and looked to the profit derived from the research. The majority asserted, without citing any supporting empirical evidence, that the volume of biological materials available for research would be reduced. In so doing, they ostensibly ignored the possibility that granting individual property rights might simply result in the need to obtain a patient's

[31] *Moore* SC, above n 12, at 493.
[32] Ibid.
[33] Ibid.
[34] Ibid, at 494.
[35] Ibid, at 493.
[36] Ibid, at 494.
[37] Ibid.

consent.[38] The majority also asserted that recognising individual property rights would dramatically reduce the economic incentive to conduct medical research. Economic incentive and certainty of title arguments cannot, by themselves, be invoked to deny a source's property rights. These desirable outcomes can also be achieved if a source is accorded property rights. For example, if it were clear that a source had property rights in their biological materials, those rights could then be transferred to a recipient by way of sale or gift. Recipients may in turn be required to compensate sources, but such compensation would not necessarily undermine the economic incentive to conduct research and develop therapeutic drugs. Indeed, disclosure to a patient that economic benefits might follow, coupled with an offer of a ten per cent share, would potentially increase the availability of materials.

Instead, the majority argued that a reduction in economic incentives would follow because of the need to share profits with patients and the uncertainty over title to biological materials once the researcher took possession of them. Empirical research conducted since *Moore* suggests that it is not clear what impact recognising a source's property rights would have on the supply of biological materials.[39] One legislative development appears to support the majority's claim. In 1995, Oregon was the first US state to grant property rights to individuals in respect of separated biological materials.[40] However, the Oregon legislature amended the law in 2001 to specify that biological materials are not property.[41] The law was amended because of the perceived potential threat to medical research.[42]

In *Moore*, Broussard J dissented in part. Although he accepted the majority's conclusion in respect of the breach of fiduciary duty, he held that Moore had a cause of action for conversion. In critically analysing the majority judges' reasoning, he observed that they left open the question of whether separated biological materials could be classified as property.[43] Broussard J noted that 'the majority's analysis . . . rests on the proposition that a patient retains no ownership interest in a body part once the body part has been removed from his or her body'.[44] Broussard J concluded that patients retained a right to control separated biological materials and that this right to control should be protected by the law of conversion.[45]

38 For an argument that the recognition of property rights may in fact increase patient participation, see: DM Gitter, 'Ownership of Human Tissue: A Proposal for Federal Recognition of Human Research Participants' Property Rights in Their Biological Material' (2004) 61 *Washington and Lee Law Review* 257, 315. Further, for a discussion of the potential benefits of a market in human organs, see: A George, 'Marketing Humanity: Should We Allow the Sale of Human Body Parts?' (2006) 7 *UTS Law Review* 11.

39 Medical Research Council, *Human Tissue and Biological Samples for Use in Research* (London, November 1999) [2.4]; Australian Law Reform Commission, *Protection of Human Genetic Information* (ALRC DP 66, Australian Government Publishing Service, Canberra, 2002) [17.36]–[17.37].

40 Oregon Genetic Privacy Act 1995, § 4(1).

41 Instead, the amended legislation (altered by Senate Bill 114 of 2001 § 15) stipulates that biological materials and information are private and must be protected.

42 A Onion 'Should You Own Your Own Genes' ABC News (US) <http://www.ukabcnews.go.com/sections/News/ geneprivacy.html> (10 May 2001).

43 *Moore* SC, above n 12, at 501.

44 Ibid.

45 Ibid, at 506.

Mosk J also dissented and delivered a forceful opinion rejecting the majority's two conclusions. First, he addressed the claim that Moore had no cause of action because he did not retain an 'ownership' interest in his cells following their removal from his body. Mosk J noted that common law authority neither supported nor rejected that proposition.[46] Next, Mosk J considered section 7054.4 of the Health and Safety Code and concluded that it did not authorise the commercial exploitation of Moore's biological materials.[47] Even if it did permit such exploitation, Mosk J asserted that it had not stripped away the proprietary nature of the materials. He observed that the concept of property referred to a 'bundle of rights' including the rights to possess, use, exclude others and dispose of things by way of sale or gift.[48] Ultimately, Mosk J held that, even if section 7054.4 limited Moore's use and disposal of his biological materials, Moore retained property rights to them.[49]

Secondly, Mosk J disputed the majority's claim that Moore had no property rights because the cell-line was factually and legally distinct from Moore's original cells. He maintained that it was irrelevant whether the cell-line was factually distinct, for the legal issue at stake was whether Moore's original biological materials were property.[50] Mosk J rejected the majority's claim that the patent settled the question. Instead, he observed that the relevant question for the court was the legal status of the cells at the moment when they were removed from Moore's body.[51]

Thirdly, Mosk J tackled the majority's opinion that the conversion cause of action should not be extended. He observed that the majority's principal policy objection was the potential 'disabling civil liability' for researchers.[52] In Mosk J's view, it did not follow that the application of the law of conversion would hinder research. He noted that medical research had become commercialised, with a resulting decline in the free access to materials that researchers traditionally enjoyed.[53] Challenging the majority's position, Mosk J outlined two policy considerations promoting recognition of an individual's property rights.[54] The first concerned the dignity of the human body, which is undermined when researchers are permitted to further their own interests without the patient's participation. The second concerned the way in which notions of fairness militate against one member of society profiting at the expense of another. In Mosk J's view, these considerations justified extending the conversion cause of action. Regrettably, despite providing forceful objections to the majority judgment, Mosk J did not consider in any detail the question of how and why property rights are created on the separation of biological materials.

[46] *Moore* SC, above n 12, at 508.
[47] Ibid.
[48] Ibid, at 509.
[49] Ibid, at 510.
[50] Ibid, at 511.
[51] Ibid.
[52] Ibid, at 513.
[53] Ibid.
[54] Ibid, at 515.

Both the Court of Appeal and the Supreme Court (in particular, the majority decisions) failed to address adequately the question of an individual's property rights to separated biological materials. This is unfortunate, particularly in light of the weight that the Supreme Court decision is accorded in other common law countries. The majority did not conclude that separated biological materials could never be the subject of property rights, but rather that a source retains no ownership interest over those materials once they are detached from their body. As the majority observed: 'we do not purport to hold that excised cells can never be property for any purpose'.[55] In doing so, the majority effectively sidestepped the fundamental question, basing the decision on the policy concerns over individuals being permitted to sell their separated biological materials. Only two subsequent US cases have considered *Moore* in any detail: *Greenberg v Miami Children's Hospital Research Institute*[56] and *Washington University v Catalona*.[57] Each is addressed in turn below.

2 *Greenberg v Miami Children's Hospital*

The *Greenberg* proceedings were brought by the parents of children affected by Canavan disease against researchers who developed and patented a test for the disorder. Canavan disease is a relatively rare, but always fatal, inherited degenerative brain disorder. In 1987, the plaintiffs approached a research physician, Dr Matalon, and requested his involvement in discovering the genes responsible for the disease so that tests could be administered to determine carriers. Over a number of years the plaintiffs located other Canavan families and convinced them to provide biological materials, familial information and financial support. The plaintiffs alleged that they had participated on the understanding that any prenatal testing developed from the research would be provided on an affordable and accessible basis. Utilising the biological materials, Dr Matalon isolated the gene responsible for the Canavan disease. Following successful trials, a test for the disease was developed to which the defendants sought to restrict access. The plaintiffs claimed they were never informed that the defendants intended to seek a patent on the research products.

Proceedings were initiated by the plaintiffs claiming: (1) lack of informed consent; (2) breach of fiduciary duty; (3) unjust enrichment; (4) fraudulent concealment; (5) conversion; and (6) misappropriation of trade secrets. The conversion claim alleged that the '[p]laintiff research subjects had a property interest in their blood, tissue, urine and autopsy samples and those of their minor children, and in

55 Ibid, at 493.

56 264 F Supp 2d 1064 (US DC Florida 2003). Although *Greenberg* was commenced in the US District Court for the Northern District of Illinois, it was transferred to the US District Court for the Southern District of Florida on 8 July 2002.

57 437 F Supp 2d 985 (USDC Ed Mo 2006) (*Catalona*); WL 1758268 (20 June 2007) (8th Cir) (*Catalona* CA).

the genetic information contained therein'.[58] In response, the defendants brought a motion to strike out the entire proceedings.

Moreno J of the Federal District Court of Florida struck out all causes of action, except the unjust enrichment claim. He considered that the plaintiffs' financial contributions were sufficient to plead a cause of action in unjust enrichment.[59] The conversion claim was dismissed, and Moreno J held that the plaintiffs did not have property rights in either the biological materials or the genetic information.[60] The judgment drew heavily on the reasoning of the Supreme Court majority in *Moore*, and Moreno J presented three justifications for dismissing the conversion claim. First, he applied *Moore* and held that an individual could not sustain a claim in conversion because a donor had no property rights once a donation had been made. For Moreno J, 'the property right in blood and tissue samples . . . evaporates once the sample is voluntarily given to a third party'.[61] This line of reasoning appears to start from the premise that an individual possesses innate property rights to their biological materials. On their donation, the rights are transferred to the third party recipient. The unresolved legal issue then becomes a question of fact in each case as to whether the property rights of each source are transferred to the defendants by the donation. Moreno J did not explore the transfer of the property rights in any detail. Rather, he assumed that a transfer occurred because the materials were voluntarily donated.

Moreno J's conclusion does not necessarily follow. For instance, the plaintiffs may have donated their biological materials for a specific purpose. Thus, in *Greenberg*, the plaintiffs alleged that the materials were donated for the purpose of discovering the gene responsible for Canavan disease and not for the purpose of developing a commercially patentable test. Arguably, in such circumstances, there is not a complete transfer of the source's property rights. Rather, the donation of biological materials may be better characterised as a conditional gift or bailment. A conditional gift is one made on condition that it be used (or not used) in a particular way. Ownership of a chattel may be transferred to the donee, but revert to the donor should specified events occur.[62] A bailment is defined as a 'possessory relationship by which a bailor transfers possession of a chattel to a bailee'.[63] Where there is no contract, the rights and duties of the bailor and the bailee may be defined according to the law of tort.[64] If the defendants in *Greenberg* acted inconsistently with the terms of the conditional gift or bailment, an action would lie. Moreno J did not consider either of these characterisations. The applicability of

[58] ND Illinois Oct 30 2000 (No OOC 6779) (copy on file with author) 21.

[59] *Greenberg*, above n 56, at 1073.

[60] Ibid, at 1076. For a discussion of the potential scope of a cause of action in unjust enrichment, see: DL Greenfield, 'Greenberg v Miami Children's Hospital: Unjust Enrichment and the Patenting of the Human Genetic Material' (2006) 15 *Annals of Health Law* 213.

[61] *Greenberg*, above n 56, at 1075.

[62] R Chambers, 'Conditional Gifts' in N Palmer and E McKendrick (eds), *Interests in Goods* (London, LLP, 2nd edn, 1998) 430.

[63] M Bridge, *Personal Property Law* (Oxford, OUP, 3rd edn, 2002) 33.

[64] N Palmer, *Bailment* (London, Sweet & Maxwell, 2nd edn, 1991) 44–45.

the doctrines of conditional gifts and bailment to separated biological materials is explored further in Part II.

Secondly, Moreno J opined that the patented research was factually and legally distinct from separated biological materials.[65] The criticisms levelled at the reasoning of the Supreme Court in *Moore*, outlined above, apply to this aspect of Moreno J's opinion. The conversion claim relates to the original biological materials, not to the subsequently developed materials and patent. The relevant question for Moreno J was the legal status of the biological materials at the moment of their removal from the children's bodies.

Finally, Moreno J emphasised the potential impact of the recognition of property rights on medical research. He concluded that recognising such property rights 'would cripple medical research as it would bestow a continuing right for donors to possess the results of any research conducted by the hospital'.[66] This justification is inconsistent with his earlier reasoning that sources abandon their property rights once their biological materials are donated to medical research. In any event, Moreno J's concern for medical research is open to the same criticism as that directed at the Supreme Court majority's position in *Moore*: no consideration was given to the possibility that researcher and patient might share the economic benefits.

Taking the above objections into account, the three justifications advanced by Moreno J are not persuasive. Each justification exposes legal ambiguities requiring further consideration. Unfortunately, these questions were not tested at an appellate level after a settlement was reached out of court.[67]

3 *Washington University v Catalona*

Catalona concerned a dispute between Washington University (the plaintiff) and Dr William Catalona (the defendant) over the ownership of biological materials donated for medical research.[68] Washington University is a leading private research university and Catalona is a respected urologist and medical researcher of prostate cancer. Catalona was employed by Washington University between 1976 and 2003, and during that period he was instrumental in establishing the GU Biorepository collection and the storage of biological materials for research. The

[65] *Greenberg*, above n 56, at 1074–75.

[66] Ibid, at 1076.

[67] The terms of the settlement are confidential, although a press release stated that the settlement provided for royalty-based genetic testing by licensed laboratories and royalty-free research by institutions and scientists researching for a cure: http://www.canavanfoundation.org/news/09-03_miami.php (accessed on 1 June 2007).

[68] After Washington University sued Catalona, a group of patients were added as necessary parties. Eight patients of Dr Catalona (Richard Ward, Thomas McGurk, Luis Garcia, Antonio Castro, Phillip Wilard, Ivan Parsons, James Ellis and Michael Missios) who had donated biological materials to the GU Biorepository, were permitted to intervene in the case and refer to themselves as 'patient/defendants'. These patients claimed that they owned their biological materials and advocated transferring them to Northwestern University.

GU Biorepository holds biological specimens of prostate tissue, blood and DNA samples for prostate cancer research. Patients of Catalona (and several other Washington University physicians) contributed biological materials for research. There were approximately 30,000 research participants enrolled in prostate cancer research studies and 2,500–3,000 of those participants had been Catalona's patients.[69] Persons who contributed samples to the GU Biorepository were referred to as 'research participants' throughout the decision. Other research institutions requested and received biological materials from the GU Biorepository for research projects outside Washington University and the material was transferred to those research institutions through Material Transfer Agreements (MTAs). Seven of the MTAs were signed by Catalona and those documents acknowledged Washington University as the owner of the biological samples.[70]

In order to participate in the GU Biorepository, research participants were required to sign 'informed consent' forms. Although there were variations in the forms used, typical features of the forms included statements that:

(a) the collection of samples was for medical research and not medical care;
(b) research participants could not 'claim ownership rights' to any medical or scientific product that resulted from the research;
(c) the word 'donate' was used to characterise the delivery of the sample from the research participant to the Washington University;
(d) by participating, the research participant made 'a free and generous gift of blood, tissue and/or DNA to research that may benefit others'; and
(e) participation was voluntary and participants could choose not to participate in the research study or withdraw consent at any time.[71]

If a research participant chose to withdraw consent, then under federal and state regulations, Washington University was permitted to: (1) destroy the sample; (2) store the sample indefinitely without using it any further in the research protocol; or (3) choose to 'anonymize' the samples and continue to use them in certain areas of research.[72]

In 2003 Catalona left Washington University and moved to Northwestern University, where he continued his prostate cancer research. Before leaving Washington University, Catalona wrote to all research participants, whether or not they had been his patients. Catalona's letter informed the research participants of his departure from Washington University and his intention to continue prostate cancer research. The letter then asked the research participant to sign an authorisation to release the biological materials they had donated to the GU Biorepository from Washington University to Catalona.[73] Approximately 6,000

[69] *Catalona*, above n 57, at 988–89.
[70] Ibid, at 989.
[71] Ibid, at 990.
[72] Ibid, at 992.
[73] Ibid, at 993.

research participants signed the form and returned it to Catalona.[74] The letter was sent without the approval of Washington University or Northwestern University. Washington University refused to release the biological materials to Catalona and commenced proceedings to prevent Catalona from moving the biological materials stored in the GU Biorepository and requested a declaratory judgment that it owned the biological materials.

The principal legal issue in the case was: once a source has made a voluntary donation of biological material for medical research to a research institution, does the source retain property rights that can be transferred to a third party?[75] Washington University argued that the sources made voluntary donations and once the gifts were delivered to Washington University, the university became the owner of those biological materials with the right to control their use and storage. Catalona argued that Washington University never owned the biological materials because the sources donated the biological materials with the intent that the materials would stay with Catalona for the purposes of his research.

Limbaugh J started his analysis by noting that Washington University had exclusive control of the biological materials and that the university had continually asserted its ownership interests in the biological materials stored in the GU Biorepository.[76] He then assessed the principles that *Moore* and *Greenberg* established in the context of medical research. According to Limbaugh J, *Moore* held that patients did not have ownership rights over excised cells and that, even if they did have some limited right to control excised cells, such a right was already protected by the doctrine of consent. *Greenberg* stood for the proposition that donors cannot claim a property interest in body tissue which is donated for medical research.[77] Limbaugh J considered that both *Moore* and *Greenberg* 'found the research participant to be a "donor" who had parted with any semblance of ownership rights once their biological materials had been excised for medical research'.[78] This statement accords with the approach that Moreno J took in *Greenberg*. It does, however, distort the Californian Supreme Court decision in *Moore*. The majority in *Moore* did not consider that the ownership interest in separated biological materials had been transferred from the source to the researcher. Instead, the majority drew heavily on policy reasons to deny Moore's ownership interest and hence the conversion claim.

Nevertheless, using *Moore* and *Greenberg* as the starting point, Limbaugh J found that the research participants were donors and that donation of the biological materials to the GU Biorepoistory constituted an *inter vivos* gift.[79] By adopting this approach, Limbaugh J developed the analysis that Moreno J had touched on in *Greenberg*. In *Greenberg*, Moreno J suggested that, although an individual

[74] Ibid.
[75] Ibid, at 994.
[76] Ibid.
[77] Ibid, at 995–96.
[78] Ibid, at 997.
[79] Ibid.

may have property rights in separated biological materials, those rights 'evaporate' once they are voluntarily given to a third party.[80] This approach presupposes that property rights initially vest in the source. It then becomes a question of whether these property rights are transferred by way of gift or other means.

In *Catalona*, Limbaugh J set out the three elements of an *inter vivos* gift: (1) present intention of the donor to make a gift; (2) delivery of the property by the donor to the donee; and (3) acceptance by the donee, whose ownership takes effect immediately and absolutely.[81] The question was whether the research participants had transferred their property rights in the biological materials to Washington University. Limbaugh J found that the research participants signed consent forms clearly stating that they agreed to participate in medical research studies at the University.[82] He found that there was no indication in any of the forms that the rights over the biological materials were being transferred to Catalona.[83] Instead, Limbaugh J found that the research participants intended to make *inter vivos* gifts in the form of donations to Washington University for medical research.[84] Limbaugh J made this finding despite the consent forms permitting research participants to withdraw their consent at any time.

In reaching this conclusion, Limbaugh J rejected the contentions of Catalona that the research participants made a bailment of their biological materials and not an *inter vivos* gift.[85] Limbaugh J noted that a bailment is made on the condition that the property must be restored to the bailor according to the bailor's directions as soon as the purpose of the bailment ceases.[86] In this case, Limbaugh J found that there was no evidence that the research participants had informed Washington University that they expected the biological materials to be returned to them.[87] Further, the 'medical research community itself has never considered the relationship between a research participant and a medical research institution to be one of bailment'.[88]

In concluding his analysis, Limbaugh J also referred to policy considerations and noted that they too supported a finding in favour of Washington University. His concerns were based on access to materials by the scientific community.[89] He considered that '[t]he integrity and utility of all biorepositories would be seriously threatened if Research Participants could move their samples from institution to institution any time they wanted. No longer could research protocols rely on aggregate collections since individual samples would come and go'.[90] These

[80] *Greenberg*, above n 56, at 1075.
[81] *Catalona*, above n 57, at 997.
[82] Ibid.
[83] Ibid.
[84] Ibid, at 998.
[85] Ibid, at 1001.
[86] Ibid.
[87] Ibid.
[88] Ibid.
[89] Ibid, at 1002.
[90] Ibid.

statements were simply assertions as Limbaugh J cited no evidence or empirical studies in support of them.

Ultimately, Limbaugh J concluded that the defendants were not entitled to injunctive relief. He held that Washington University 'owns all biological materials, including but not limited to blood, tissue, and DNA samples, in the GU Biorepository'.[91] There was no discussion by Limbaugh J of the work or skill principle as creating the property rights. Limbaugh J's approach appeared to rest on the assumption that property rights are created on the detachment of the biological materials: that is, because the research participants transferred property rights to Washington University by way of gift, then the property rights must have initially vested in the research participants. The act of detachment itself may therefore constitute a second event whereby the law recognises that property rights can be created in respect of separated biological materials.

Catalona appealed to the United States Court of Appeals for the Eighth Circuit.[92] The Eighth Circuit (Wollman, Riley and Shepherd JJ) considered that the principal issue was whether individuals who 'make an informed decision to contribute their biological materials voluntarily to a particular research institution for the purpose of medical research retain an ownership interest allowing the individuals to direct or authorize the transfer of such materials to a third party'.[93] The court concluded that the research participants had retained no such rights and had instead gifted their biological materials to the GU Biorepoistory. Thus, the rights the participants retained were not of a proprietary nature. To this end, the court observed:

> Viewing these provisions in their entirety, it is evident the RPs did not retain the right to revoke and physically repossess the donated biological materials. Nor did the RPs retain the right to direct or authorize the use, transfer, or destination of the biological materials after their donation. The [research participants'] subsequent rights to their biological materials were expressly limited to the option to discontinue participation in the study to avoid answering additional questions, donating more biological materials, or allowing their biological materials to be used for further research. Beyond these particular and limited rights, the [research participants] retained no greater interest with regard to their biological materials. Such rights cannot be equated with or interpreted to include the broad privileges or proprietary interests advocated by the defendants.[94]

Despite this finding, the court did not explicitly address the question of how property rights over the biological materials were initially created and allocated. Indeed, the court did not state that the research participants owned or even could own separated biological materials. Nevertheless, the reasoning of the Eighth Circuit appears to be based on the same assumption adopted by Limbaugh J: that property rights are created on the detachment of the biological materials from the research participant and then allocated to the research participant. If the property

91 Ibid.
92 *Catalona* CA, above n 57.
93 Ibid, at 9.
94 Ibid, at 12.

rights to the biological materials did not initially vest in the research participants, those research participants could not have transferred the ownership of the materials to Washington University by way of gift. In embracing this approach, the Eighth Circuit did not consider or even mention the decision of the Californian Supreme Court in *Moore*. This is surprising given that *Moore* (at least on one view) is inconsistent with the reasoning of the Eighth Circuit. The lack of analysis is particularly unfortunate given the weight that common law jurisdictions have placed on *Moore*.

4 Summary

Two decisions since *Moore*—*Catalona*, and to a lesser extent, *Greenberg*—suggest that property rights in separated biological materials are created on detachment, and that those rights can be transferred from the source to a third party on their donation for medical research. Nevertheless, the leading US authority remains *Moore*, which rejected the proposition that a living person can claim property rights to separated biological materials. In that case, the Californian Supreme Court did not adequately address the more fundamental question as to why a living person cannot claim property rights in separated biological materials. Instead, the decision focused on the policy impact of individuals being permitted to sell separated biological materials. Such a policy analysis should not be the first enquiry. The more fundamental questions are: (1) who is entitled to control separated biological materials? and (2) does property law apply to determine if property rights are created over those materials? Only once these questions have been addressed, can and should policy-based issues be considered in determining the content of any property rights.

C Property Rights

It is an open question in English law as to whether property rights are created on the separation of biological materials. Despite the pervading influence of *Moore*, there are suggestions in statute and at common law that separated bodily materials can be classified as property.[95] These statutory regimes and common law cases are analysed here from the perspective of five different types of bodily materials separated from a living person: (1) biological materials; (2) cell-lines; (3) gametes; (4) hair clippings; and (5) waste products.

[95] The fact that Germany, as a civil law jurisdiction, recognises property rights in separated biological materials has already been noted. See, above n 1, and accompanying text.

1 Biological Materials

(a) Statute

There is no current legislation in England that specifically determines the legal status or allocation of property rights in relation to biological materials. Chapter four analyses in detail the effect of the most recent legislative development in the common law world—the Human Tissue Act 2004 (HTA 2004). This legislation does not, however, vest property rights in individuals who have bodily materials separated from them.

In England, the Human Organ Transplants Act 1989 UK (Transplants Act) contained a statutory prohibition against trafficking in transplantable organs.[96] This statutory prohibition has been transposed into the HTA 2004[97] and is examined more fully in chapter four. Scotland has recently implemented a similar statutory prohibition.[98] In Australia, the state human tissue legislation also contains a statutory prohibition against the sale of human organs.[99] It should be noted that in some Australian states the statutory prohibition does not apply if the tissue has been subjected to processing or treatment, which is a statutory reflection of the work or skill exception.[100] In the US, the National Organ Transplant Act is the federal law that prohibits the sale of human organs for use in human transplantation.[101] American states have also prohibited commercial dealings in human organs.[102] One author has suggested that the existence of statutes prohibiting commercial dealings in organs 'portrays the human body as an article of commerce'.[103] However no statutory reference is made to human organs as property in this English, Australian and US legislation. These legislative regimes cannot, therefore, be invoked in support of individual property rights in human organs.

[96] Transplants Act, s 1.

[97] HTA 2004, ss 33 and 34.

[98] Human Tissue (Scotland) Act 2006 asp 4, s 20.

[99] Human Tissue Act 1983 (NSW) s 32; Human Tissue Act (Vic) s 38; Transplantation and Anatomy Act 1979 (Qld) s 40; Transplantation and Anatomy Act 1983 (SA) s 35; Human Tissue and Transplant Act 1982 (WA) s 29; Human Tissue Act 1985 (Tas) s 27; Transplantation and Anatomy Act 1978 (ACT) s 44; Human Tissue Transplant Act 1979 (NT) s 24.

[100] See, for example, s 32(2) Human Tissue Act 1983 (NSW): '[s]ubsection (1) does not apply to or in respect of the sale or supply of tissue if the tissue has been subjected to processing or treatment and the sale or supply is made for the purposes of enabling the tissue to be used for therapeutic purposes, medical purposes or scientific purposes'. See, also: Transplantation and Anatomy Act 1983 (SA) s 35(3); Human Tissue Act 1985 (Tas) s 27(2); Transplantation and Anatomy Act 1978 (ACT) s 44(2); Human Tissue Transplant Act 1979 (NT) s 24(4).

[101] National Organ Transplant Act 42 USCA (2000) § 274e(a).

[102] The Uniform Anatomical Gift Act (UAGA) was approved by the National Conference of Commissioners on Uniform State Laws in 1968 and has since been adopted in all 50 states. The UAGA prohibits the purchase or sale of organs for use in transplantation or therapy: § 10(a). For a discussion of the state laws, see: EB Seeney, '*Moore* 10 Years Later—Still Trying to Fill the Gap: Creating a Personal Property Right in Genetic Material' (1998) 32 *New England Law Review* 1131, 1153–56.

[103] R Rao, 'Property, Privacy and the Human Body' (2000) 80 *Boston University Law Review* 359, 376.

In considering English statute law, two European instruments also warrant examination. Under the EU Council Directive on the Legal Protection of Biotechnological Inventions (Biotech Directive)[104]—which has been implemented in English law[105]—isolated DNA may be patentable in specified circumstances, even if the material is structurally identical to material found in nature. Thus, the act of isolating biological materials from their surroundings, or producing them in an artificial way, is sufficient to place the materials so isolated or so produced within the realm of inventions. No reference is made in the Biotech Directive to any interests that individuals may have in biological materials obtained from them. Instead, the Biotech Directive provides that researchers must ensure that all human donors of biological materials provide fully informed consent to the removal of such materials.[106] The recitals to the Biotech Directive do not address the legal status of biological materials or whether property rights exist in respect of them.

More recently, EU Member States agreed to the implementation of a common regulatory framework (again, set out in a Council Directive) to ensure the safety of cells and tissues that are transplanted into, or onto, the human body (Tissue Directive).[107] Although the use of organs raises similar issues, they are excluded from the Tissue Directive.[108] Essentially, the Tissue Directive provides that donated and processed biological materials must be tested for infection and that donors should not be remunerated because of safety concerns.[109] Importantly, the Tissue Directive establishes that the consent of each donor is required for the use of biological materials.[110] It does not, however, make any determination as to the legal status of biological materials.

Similarly, Australian legislation does not determine the legal status of biological materials. The Australian Law Reform Commission recently recommended that 'legislation should not be enacted to confer fully proprietary rights in human genetic samples',[111] and this recommendation was accepted by the Australian Federal Government.[112] Existing legislation in Australia instead requires an indi-

[104] Council Directive (EC) 98/44 on the Legal Protection of Biotechnological Inventions [Legal Protection of Biotechnological Inventions] [1998] OJ L213/13.

[105] Patents Regs 2000 (SI 2000/2037).

[106] Council Directive (EC) 98/44, recital 26.

[107] Council Directive (EC) 2004/23 on Setting Standards of Quality and Safety of the Donation, Procurement, Testing, Processing, Preservation, Storage and Distribution of Human Tissues and Cells [2004] OJ L/102/48. This directive has been implemented in the United Kingdom through The Human Tissue (Quality and Safety for Human Application) Regulations 2007 (SI 2007/1523), which came into force on 5 July 2007.

[108] Council Directive (EC) 2004/23, recital 9.

[109] Council Directive (EC) 2004/23, recitals 18 and 19.

[110] Council Directive (EC) 2004/23, art 13.

[111] Australian Law Reform Commission, *Essentially Yours: The Protection of Human Genetic Information* (ALRC 96 Australian Government Publishing Service, Canberra, 2003) recommendation 20–1.

[112] Government Response to Recommendations of Australian Law Reform Commission, *Essentially Yours: The Protection of Human Genetic Information* (Commonwealth Government, Canberra, 2005) 23. See, also: Privacy Legislation Amendment Act 2006 (Cth).

vidual's consent for the donation of regenerative tissue.[113] Regenerative tissue is defined as any tissue that can be replaced by the body after removal.[114] The level of protection has been extended in New South Wales following amendments contained in the Human Tissue and Anatomy Legislation Amendment Act 2003 (NSW).[115] Section 34(1)(b) of the Human Tissue Act 1983 (NSW) provides that tissue removed during medical, surgical or dental procedures is not to be used for other purposes without written consent. This requirement for consent is undermined by section 34(1)(b1) of the Human Tissue Act 1983 (NSW) which provides that no consent is required for the retention and use for therapeutic, medical or scientific purposes of 'small samples of any tissue' that is lawfully removed and retained in the form of a tissue slide or block. Although these provisions do not confer property rights on individuals, section 34(1)(b) represents legislative recognition that individuals retain a right to control separated biological materials.

A different position regarding the statutory recognition of property rights to separated biological materials is found in the US. Various legislatures in the US have considered allocating individuals property rights to biological materials removed from them. In 1995, a model statutory instrument was proposed—section 104(a) of the Genetic Privacy Act (GPA)[116]—that recognised individual property rights to such materials. Section 104(a) provided that an individually identifiable DNA sample is the property of its source. The US Congress introduced a version of the GPA in 1995. Although it identified a DNA sample as property, this version of the GPA did not assign those property rights to anyone,[117] and was never enacted.[118] Moreover, this chapter has already referred to the Oregon legislation, which provided an example of a statutory creation of individual property rights.[119]

Some US states have also sought to protect the use of genetic information by declaring that genetic information is the personal property of an individual.[120]

[113] Human Tissue Act 1983 (NSW) ss 7, 9, 12; Transplantation and Anatomy Act 1979 (Qld) ss 10, 12, 13; Transplantation and Anatomy Act 1983 (SA) ss 9, 15; Human Tissue Act 1985 (Tas) ss 7, 9, 14; Human Tissue Act 1982 (Vic) ss 7, 9, 10; Human Tissue and Transplant Act 1982 (WA) ss 8, 15; Transplantation and Anatomy Act 1978 (ACT) ss 8, 10, 15; Human Tissue Transplant Act 1979 (NT) ss 8, 10, 11.

[114] Human Tissue Act 1983 (NSW) s 4(1); Transplantation and Anatomy Act 1979 (Qld) s 5(1); Transplantation and Anatomy Act 1983 (SA) s 5(1); Human Tissue Act 1985 (Tas) s 3(1); Human Tissue Act 1982 (Vic) s 3(1); Human Tissue and Transplant Act 1982 (WA) s 3(1); Transplantation and Anatomy Act 1978 (ACT) s 4(1); Human Tissue Transplant Act 1979 (NT) s 4(1).

[115] These amendments were made following an inquiry into practices at the New South Wales Institute of Forensic Medicine in Sydney: B Walker SC, *Inquiry into Matters Arising from the Post-mortem and Anatomical Examination Practices of Forensic Medicine* (Government of the State of NSW, Sydney, 2001).

[116] This is a Model Bill and has not passed through the US Congress: P Roche, L Glantz and G Annas, 'The Genetic Privacy Act: A Proposal for National Legislation' (1996) 31 *Jurimetrics* 1, 1.

[117] For a discussion of this Federal Bill, see: MJ Lin, 'Conferring a Federal Property Right in Genetic Material: Stepping into the Future with the Genetic Privacy Act' (1996) 22 *American Journal of Law and Medicine* 109.

[118] Genetic Non Discrimination and Confidentiality Bill (Bill No S422, introduced 11 March 1997).

[119] Oregon Genetic Privacy Act 1995, § 4(1).

[120] Florida (Florida Statutes Title XLIV Chapter 760 § 760.40); Georgia (Code § 33-54-1); Louisiana (Acts No 1418 § 213.7); Colorado (Colorado Statutes § 10-31104.7).

These legislative enactments have a different focus: they create property rights in the information once it has been extracted from the biological materials. For example, the relevant Colorado statute provides that '[g]enetic information is the unique property of the individual to whom the information pertains'.[121] One construction of the Colorado statute presupposes that biological materials are also the personal property of the individual. This is not a persuasive interpretation of the statute because, on its terms, the legislation does not regulate the physical materials, but instead only the genetic information extracted from them.

(b) Common Law

There is limited English common law authority considering the legal status of biological materials. In *R v Rothery*,[122] the defendant was charged with theft after he removed a blood sample contained in a capsule from a police station.[123] At trial, he was convicted of theft and the statutory offence under section 9(3) of the Road Traffic Act 1972 of failing to supply a specimen. On appeal, the court was asked to determine whether the defendant should have been charged with the statutory offence. In considering the statutory offence, Scarman LJ observed that 'he must have provided the police officer with the specimen in the sense of putting him in possession or control of it, otherwise he could not have stolen it from him under section 5(1) of the Theft Act 1968'.[124] As a result, Scarman LJ quashed the conviction for the statutory offence.

There are two possible interpretations of the legal status that Scarman LJ implicitly accorded to the blood sample. Although he did not openly hold that the blood specimen was property, Scarman LJ's language suggests that it did constitute property capable of being stolen. On the other hand, the description of the charge suggests that it was the theft of the 'capsule' that the court proceeded to discuss, rather than its contents *per se*. If this latter view is correct, then there was no need for Scarman LJ to consider whether the blood itself could be property. The lack of clarity casts doubt over the case as authority for either interpretation.

Similarly, there is limited common law authority on this matter in Australia. In 1977, the Australian Law Reform Commission (ALRC) concluded in its *Human Tissue Transplants Report* that the common law was silent on the legal status of biological materials.[125] More recently, in 2003 the Australian Law Reform Commission released *Essentially Yours: The Protection of Human Genetic Information* and noted that 'the common law position on property rights in human tissue samples is not well developed, and there is no clear judicial statement on the issue'.[126]

[121] Colorado Statutes, § 10-31104.7.
[122] [1976] RTR 550 (CA).
[123] *Rothery*, ibid at 553.
[124] Ibid, at 552–53.
[125] Australian Law Reform Commission, *Human Tissue Transplants* (ALRC 7, Australian Government Publishing Service, Canberra, 1977) 7.
[126] *Essentially Yours*, above n 111, at [20.10].

The most extensive judicial discussion in Australian common law is that of Master Sanderson of the Western Australian Supreme Court in *Roche v Douglas.*[127] In *Roche*, the plaintiff brought an action under the provisions of the Inheritance (Family and Dependants Provision) Act 1972 (WA) (Inheritance Act) relating to the will of Edward Rowan (the deceased). The deceased died in 1999 and the plaintiff claimed that he was her natural father and that she was entitled to make a claim under section 7(1)(c) of the Inheritance Act. The executor of the will (the defendant) denied the claim.

The deceased had undergone surgery at a hospital. Certain biological materials were taken from him, preserved and held at a pathology laboratory. The plaintiff sought an order seeking access to the materials so that she could submit them for DNA parentage testing.[128] The defendant argued that the samples in question were not 'property', and so did not satisfy the threshold requirement for an order to be made under the Supreme Court Rules.

The central issue in *Roche* was, thus, whether the biological materials could be classified as property. Master Sanderson concluded in the affirmative. He considered that the line of authority establishing the 'no property' principle in respect of cadavers did not assist the court. Instead, Master Sanderson drew on Griffith CJ's observation in *Doodeward v Spence*[129] regarding the necessity to apply principles of law in line with reason and good sense. On the facts, he considered that there was a compelling reason to classify the materials as property, namely savings in time and cost.[130] Further, he considered that there was no logical justification to ignore the physical presence of the biological materials.[131] To do otherwise, he opined, would 'create a legal fiction'.[132]

Roche is not strong authority, for it is a decision at a preliminary hearing and, perhaps more significantly, the question of title was left unanswered, as the Supreme Court Rules did not require an order as to ownership of the property.[133] Indeed, Master Sanderson noted that property rights might have been allocated to the defendant, the hospital or the laboratory. In the event, Master Sanderson made the order against the laboratory as the party in possession of the materials. Despite the relatively weak analysis contained in both *Rothery* and *Roche*, these cases support the classification of separated biological materials as property which is not dependent on the application of work or skill. Instead, the classification appears to rest on the detachment of the biological materials.

The question of property rights has also arisen over the application of 'sale of goods' warranties to donated blood. There is no English authority discussing the question in the context of the Sale of Goods Act 1979 UK. In Australia this issue

[127] (2000) 22 WAR 331 (WA SC).

[128] Order 52 r 3(1) of the Supreme Court Rules 1971 (WA) provides that a court may order the taking of a sample of property to make a proper determination of a cause of action.

[129] (1908) 6 CLR 406 (HCA).

[130] *Roche*, above n 127, at 338.

[131] Ibid, at 338–39.

[132] Ibid, at 339.

[133] Ibid.

arose after several hundred people contracted the human immunodeficiency virus (HIV) from contaminated blood transfusions. At the time, HIV antibody screening had not been introduced. Actions were subsequently commenced against the Red Cross and relevant hospitals alleging negligence and contravention of the consumer protection provisions of the Trade Practices Act 1974 (Cth) (TPA). *E v Australian Red Cross Society*[134] served as a test case. The applicant had received a blood transfusion through which he alleged he was infected with HIV. He argued that the defendants breached the implied conditions and warranties regarding merchantability and fitness for purpose under sections 71 and 74 of the TPA. In respect of these later submissions, the applicant had to show that the defendants were 'trading corporations' within the meaning of the TPA. This required establishing that blood fell within the statutory definition of 'goods'.

At first instance, Wilcox J left this question open by holding that the contract between the applicant and Prince Alfred Hospital was for the supply of nursing services and not for the supply of blood.[135] Without a contract for the supply of goods, Wilcox J dismissed the cause of action pleaded under the TPA. Further, the applicant's appeal was dismissed by the Full Court of the Federal Court (Lockhart, Sheppard and Pincus JJ).[136] The Full Court held that there was no relevant contract for the supply of goods under section 71 of the TPA because the essence of the contract between the appellant and the hospital was for services rendered.[137] The Full Court also dismissed the negligence claim.

A similar issue arose in *PQ v Australian Red Cross Society*.[138] In that case, the plaintiff brought an action against the Australian Red Cross for negligence and breaches of sections 71 and 74 of the TPA. The plaintiff had received blood products and he became infected with HIV. In relation to the TPA claim, McGarvie J observed that the issue was whether the claim was statute-barred and concluded that it was. On this basis, McGarvie J avoided the issue of whether blood products were 'goods' for the purposes of the legislation. Neither *E v Australian Red Cross Society* nor *PQ v Australian Red Cross Society*, therefore, provides judicial authority for the proposition that blood removed from a living person can be classified as goods capable of being the subject of property rights.

The US has a more substantial corpus of jurisprudence on this issue. The legal status of blood in the US has been tested judicially—first, in claims that warranty requirements for the sale of goods apply to the supply of blood, and secondly, in claims that income which individuals derive from the sale of their blood is taxable. For warranty requirements to attach to transactions under US legislation, it is necessary to show that the provision of blood and blood products amounts to a sale of goods. US courts have, nevertheless, disagreed over how the provision of

[134] (1991) 27 FCR 310 (FCA).
[135] *Australian Red Cross Society*, ibid at 353.
[136] *Australian Red Cross Society* (1991) 105 ALR 53 (FC FCA).
[137] *Australian Red Cross Society* FC, ibid at 57.
[138] [1992] 1 VR 19 (Vic SC).

blood and blood products should be characterised. Some have held it to be a sale of goods; others the provision of a service.

In *Perlmutter v Beth David Hospital*,[139] a patient (the plaintiff) received medical treatment in a hospital where she paid for hospital services and a separate sum for the blood that she received during a transfusion. The plaintiff claimed that she received contaminated blood. She sought recovery on the basis that supplying blood constituted a sale within the Sales Act and that, therefore, implied warranties were imposed by section 96(1) and (2) of the Personal Property Law. A majority of the Court of Appeals of New York (Lewis CJ, Desmund, Van Voorhis, Fuld JJ) held that a blood transfusion performed by a hospital, in the course of medical treatment, was not a 'sale' within the meaning of warranty law. Instead, the court considered that the transfer of blood was part of the hospital's provision of a service.[140] With no separate sale of blood, the implied warranties were held to be inapplicable. *Perlmutter* has been followed by a number of other states, including, for example, in *Balkowitsch v Minneapolis War Memorial Blood Bank*,[141] *Dibblee v Dr WH Groves Latter-Day Saints Hospital*[142] and *Roberts v Suburban Hospital Association Inc.*[143] In *Balkowitsch*, the Minnesota Supreme Court held that a transaction between a non-commercial blood bank and a hospital was not a sale. The court reasoned that a non-profit corporation should be treated no differently from a hospital.[144]

Despite these examples, other US state courts have found that blood and blood products can be 'goods' with property characteristics.[145] In *Reilly v King County Central Blood Bank*,[146] the appellant claimed damages for an attack of hepatitis that she allegedly suffered as the result of blood transfusions. The appellant argued that the respondent was liable, on a theory of strict liability in tort, for the supply of defective goods. The respondent was a non-profit corporation that operated a blood bank to supply blood to those in need. Donors to the bank received no monetary compensation. The respondent organised the packaging, storing and delivery of the blood to the patient's hospital and charged patients a fee for each unit of blood used. The State of Washington Court of Appeals concluded that '[t]he transaction in this case has all the attributes of a sale. There was a transfer of property through the mutual consent of competent parties for a consideration in money paid'.[147]

Additional decisions from Florida and New York courts appear to draw a distinction between commercial blood banks and hospitals. In *Community Blood*

[139] 123 NE 2d 792 (NY App 1954).

[140] *Perlmutter*, ibid at 796.

[141] 123 NW 2d 805 (Minn SC 1965).

[142] 364 P 2d 1085 (Ut SC 1961).

[143] 532 A 2d 1081 (Md AC 1987).

[144] *Balkowitsch*, above n 141, at 810.

[145] See, for example: *Reilly v King County Central Blood Bank* 492 P 2d 246 (Wash App 1972); *Cunningham v MacNeal Memorial Hospital* 266 NE 2d 897 (Ill SC 1970).

[146] 492 P 2d 246 (Wash App 1972).

[147] *Reilly*, ibid at 248.

Bank Inc v Russell,[148] the Supreme Court of Florida held that, where a commercial bank supplied blood to a patient for consideration, it constituted a sale, and was therefore subject to a breach of warranty claim.[149] In *Carter v Inter-Faith Hospital of Queens*,[150] the New York Supreme Court considered the question in more detail. There, the plaintiff had received blood transfusions and allegedly became infected with hepatitis as a result of the transfusions. The plaintiff contended that the blood was purchased by the first defendant from National Blood Bank Inc and brought an action to recover damages for injuries suffered because of alleged negligence and breach of warranty against the hospital (the first defendant) and the National Blood Bank (the second defendant).

The court dismissed the action against the first defendant, holding that hospitals could not be liable for the breach of implied or express warranties of fitness.[151] It then considered the claim against the second defendant and whether a commercial blood bank could be liable for breach of warranty where it has transferred blood to a hospital for consideration.[152] The court opined that *Perlmutter* stood for the proposition that the supply of blood by a hospital to a patient is incidental to the services rendered and does not, therefore, constitute a sale. In *Carter*, the court found that there had been a simple transfer of blood and that there appeared to be a cause of action for breach of warranty against the second defendant.[153]

This body of US case law supports the creation of property rights to blood once it is removed from a living person. Nonetheless, even in these cases courts have not been required to determine if an individual can claim immediate property rights to blood following its removal.

In a different context, US courts have characterised money earned by individuals from the sale of their own blood as taxable income. In *United States v Garber*,[154] the defendant was charged and convicted of income tax evasion for failure to report funds received by her from the sale of her blood plasma. The defendant had a rare antibody and entered into a range of contracts over three years for her blood plasma. The court held that the defendant was liable to pay tax on the income as it considered that 'blood plasma, like a chicken's eggs, a sheep's wool, or like any saleable part of the human body, is tangible property' within the terms of the tax legislation.[155] Ultimately, the conviction was quashed because the defendant had not wilfully engaged in tax avoidance. *Green v Commissioner*[156] also considered the question of whether the continuous sale of blood could be classified as taxable income. The petitioner earned her living by selling her rare, AB-negative blood, which she had done on a regular basis for over seven years. The US Tax Court held

[148] 196 So 2d 115 (Flo Sup 1967).
[149] *Russell*, ibid at 120.
[150] 304 NYS 2d 97 (NY SC 1969).
[151] *Carter*, ibid at 98.
[152] Ibid, at 99.
[153] Ibid, at 102.
[154] 607 F 2d 92 (5th Cir 1979).
[155] *Garber*, ibid at 97.
[156] 74 TC 1229 (US Tax Ct 1980).

that the income she received from these activities was taxable income and observed that type AB-negative blood was a tangible product like any other 'raw product of nature'.[157]

In *Garber* and *Green* the courts were not required to define the legal status of separated blood, but simply determined whether income derived from its sale comprised taxable income. Potentially then, the income that individuals receive for their blood in the US could be viewed merely as payment for the provision of a service.

(c) Summary

Common law decisions such as *Rothery* and *Roche* have classified separated biological materials as property. In these cases, the courts were not required to determine the more difficult question of whether property rights in these materials could be allocated to a living person. Instead, the legal status of biological materials arose as a matter of statutory construction. In the US, there is a body of jurisprudence that classifies blood and blood products as 'goods' in two situations. The first is where a commercial blood bank provides blood or blood products for a fee; the second is where individuals sell their own blood as a source of income. Again, though, these cases have not had to determine whether a living person can claim property rights in their own blood. Together, the decisions discussed suggest that property rights can be created on the detachment of biological materials. The potential for 'detachment', as a principle explaining the creation of property rights in biological materials, is developed more fully in Part II.

2 Cell-lines

Only one common law decision, *US v Arora*,[158] has specifically considered the proprietary status of a cell-line. In *Arora*, the National Institute of Health (NIH) employed two scientists, Dr Arora and Dr Sei, to research and develop a new line of cells. Over time their relationship deteriorated and Sei found that a cell-line he created had been damaged without explanation. The US Government brought a civil action for conversion against Arora. Messitte J found, as a question of fact, that Arora had tampered with and caused the death of the new cell-line.[159] He then considered whether the cell-line was property capable of conversion. Messitte J observed that US courts had not hitherto recognised a claim for conversion over cell-lines, citing both *Moore* and *Miles v Scripps*.[160] However, he observed that in *Moore*, while the claimant had been denied property rights, Broussard J had in dissent suggested that the tort of conversion would lie where another medical centre or drug company had stolen the cells in question.[161]

[157] *Green*, ibid at 1234.
[158] 806 F Supp 1091 (Md DC 1994).
[159] *Arora*, ibid at 1097.
[160] 810 F Supp 1091 (Cal DC 1993).
[161] *Arora*, above n 158, at 1098.

In *Arora*, Messitte J considered that neither contract nor patent law applied to the wrongdoing. He found himself having to consider the legal status of the cell-line. Messitte J held it was a chattel which was capable of being converted and observed: 'if such a cause of action is not recognized, it is hard to conceive what civil remedy would ever lie to recover a cell-line that might be stolen or destroyed'.[162] Cell-lines raise more complicated questions than original biological materials, for they often comprise different physical materials. The more difficult conflict, as a matter of legal principle, arises between the individual who donates the original biological materials and the researcher who develops the cell-line. Messitte J was not required to consider this potential conflict in *Arora*. The difficult questions raised by such a conflict are considered in chapter six. That chapter examines what property principles can be applied to determine the legal status and competing rights pertaining to cell-lines.[163]

3 Gametes

(a) Statute

The strongest suggestion of property rights subsisting with respect to separated bodily materials in England is located in the Human Fertilisation and Embryology Act 1990 UK (HFEA). The HFEA was passed, *inter alia*, to regulate the creation of human embryos outside the body (*in vitro*) by means of a statutory licensing authority. It provides that the creation and storage of an embryo and the storage and use of gametes[164] can be carried out only under licences granted by the Human Fertilisation and Embryology Authority (Authority).[165] In combination, the HFEA and HFEA Code of Practice set up a detailed regulatory regime. The relevant provisions for this book concern the rights of individuals who donate gametes (gamete providers).

Three features of the HFEA are relevant. First, consent is the central principle that underpins the regime.[166] Individuals must consent to the storage and use of their gametes.[167] Further, gamete providers can specify conditions for the use of their gametes and (within limits) vary or withdraw their consent.[168] The ability to stipulate conditions gives gamete providers a continuing interest in respect of their gametes, and any resulting embryo, after donation. However, because the HFEA does not define the legal nature of the gamete providers' interest, it cannot deal with a dispute arising outside the regulatory framework. Jackson has observed that

[162] *Arora*, above n 158, at 1099.

[163] See nn 118–40 of ch 6 and accompanying text.

[164] A 'gamete' is defined in this book as either the male sex or reproductive cells (spermatozoa) or the female sex or reproductive cells (ova).

[165] HFEA, ss 3, 4.

[166] *Leeds Teaching Hospitals NHS Trust v A* [2003] EWHC 259; [2003] 1 FLR 1091 (QB) 1100.

[167] HFEA, sch 3.

[168] Human Fertilisation and Embryology Authority, *Code of Practice* (London, HFEA, 6th edn, 2003) [6.8].

the donor's right to use and dispose of their gametes 'looks very like ownership'.[169] The better view is that gamete providers have a limited right to use and dispose of their gametes. The HFEA makes no reference to individual property rights.

Secondly, the HFEA does not prohibit payment for the donation of gametes.[170] The Authority issued a direction in 1998 permitting payment on the basis that removing payment 'would seriously jeopardize the supply of sperm donors'.[171] Under the most recent direction, issued by the Authority in April 2006, gamete providers can claim compensation for all 'reasonable expenses' and 'earnings lost' when donating.[172] These payments are intended to recognise the value of the donation to compensate for donors' expenses and inconvenience.[173] Payment, in this context, is viewed as payment for a service rather than for the sale of the gamete.[174]

Thirdly, the HFEA sets out what rights gamete providers have in respect of frozen embryos. This book does not evaluate the moral status of human embryos, or the morality of their use in research, but instead asks what statutory rights individuals are granted. Under the HFEA, individuals are entitled to withdraw their consent regarding the use or storage of embryos created using their gametes.[175] Once consent is withdrawn the embryos must be destroyed. The scope of an individual's control was tested recently in *Evans v Amicus Healthcare Ltd; Midland Fertility Services v Hadley*.[176] The case concerned the use and storage of frozen embryos created during relationships that had since ended. Both claimants (Evans and Hadley) sought *in vitro* fertilisation (IVF) treatment with their respective male partners. Under paragraph 6(3) of Schedule 3 to the HFEA, an embryo created by IVF could not be used for any purpose without effective consent from each person (male and female) whose gametes were used. Each man had signed a form by which he consented to the use of embryos fertilised with his sperm for the 'treatment of himself, together with his partner'. Embryos subsequently were created and stored, but the two claimants then separated from their partners. The question before the court was whether the claimants were still entitled to use the frozen embryos. Their former partners refused their consent at this stage and sought an order for the destruction of the embryos.

[169] E Jackson, *Regulating Reproduction* (Oxford, Hart Publishing, 2001) 234.

[170] Human Fertilisation and Embryology Act 1990 UK, s 12(e). That section provides that: 'no money or other benefit shall be given or received in respect of supply of gametes or embryos unless authorised by directions'.

[171] HFEA, *Directions Given Under the Human Fertilisation and Embryology Act: Giving and receiving money or other benefits in respect of any supply of gametes or embryos* Ref D 1998/1 (7 December 1998) (available at http://www.hfea.gov.uk, accessed on 1 June 2007).

[172] HFEA, *Directions Given Under the Human Fertilisation and Embryology Act: Giving and receiving money or other benefits in respect of any supply of gametes or embryos* Ref D 2006/1 (1 April 2006) (available at http://www.hfea.gov.uk, accessed on 1 June 2007).

[173] HFEA, *Code of Practice*, above n 168, [4.26].

[174] B Dickens, 'Donation and Transplantation of Organs and Tissues' in A Grubb (ed), *Principles of Medical Law* (Oxford, OUP, 2nd edn, 2004) 1041.

[175] HFEA, sch 3 [2(4)].

[176] [2003] EWHC 2161; [2004] 2 WLR 713 (Fam) (*Evans*) and [2004] EWCA Civ 727; [2005] Fam 1 (CA) (*Evans* CA).

At first instance, Wall J held that the original consents to IVF treatment were for 'treatment together' as a couple, and because the relationships no longer existed the consents were no longer effective.[177] He also held that, even if the original consents were effective, the men had validly exercised their right to withdraw consent.[178] Evans appealed and the Court of Appeal upheld Wall J's decision. The court found that the men were entitled under the HFEA to withdraw their consent under the HFEA.[179] The Court of Appeal did not, however, categorise this right as proprietary.[180]

Evans then brought proceedings in the ECtHR claiming, *inter alia*, that schedule 3 of the HFEA Act, which permitted the withdrawal of consent, violated her rights to respect for private and family life under Article 8 of the ECHR. A majority of the ECtHR[181] held that 'private life' encompasses the right to respect for both the decisions to become, and not to become, a parent and that the UK legislation did not exceed the margin of appreciation for the application of that right.[182] Evans appealed to the Grand Chamber of the ECtHR and a majority of the Grand Chamber upheld the decision of the ECtHR and found that schedule 3 of the HFEA Act did not violate Article 8.[183]

Evans illustrates that the HFEA confers on gamete providers the power to decide what should be done with their stored gametes (and any resulting embryos). The question of whether such powers constitute property rights was not tested in *Evans* or in the earlier case concerning Ms Blood and her dispute with the Authority over her application to access her dead husband's sperm in order to become pregnant through IVF.[184]

The Nuffield Council on Bioethics has suggested that the HFEA provisions endorse a property-based approach. Although the statute relies on a scheme of consents to avoid the need to decide the issue of property, nevertheless 'it contemplates that the control and disposal of gametes and embryos rest with the donor(s) and allows for the transfer of the reproductive materials between those having a licence to deal with them'.[185] Similarly, Matthews has observed that the

[177] *Evans*, ibid at [149].

[178] Ibid, at [165].

[179] *Evans* CA, above n 176, at [41] (Thorpe LJ and Sedley LJ), [100] (Arden J).

[180] Leave to appeal to the House of Lords was refused on 29 November 2004.

[181] President Casadevall, Judges Bratza, Pellonpää, Maruste, Sikuta. Judges Traja and Mijovic dissented.

[182] *Case of Evans v The United Kingdom* (2006) 43 EHRR 21, [57], [69] (President Casadevall, Judges Bratza, Pellonpää, Maruste, Sikuta). For a discussion of the ECtHR decision, see: T Annett, 'Balancing Competing Interests over Frozen Embryos: The Judgment of Solomon' (2006) 14 *Medical Law Review* 425.

[183] *Case of Evans v The United Kingdom* [2007] ECHR 264, [90]–[92] (President Rozakis, Judges Costa, Bratza, Zupančič, Lorenzen, Butkevych, Vajić, Baka, Kovler, Zagrebelsky, Mularoni, Jaeger, Björgvinsson and Ziemele). Judges Türmen, Tsatsa-Nikolovska, Spielmann and Ziemele delivered a joint dissenting opinion.

[184] *R v Human Fertilisation and Embryology Authority; Ex Parte Blood* [1997] 2 FLR 742 (CA).

[185] Nuffield Council on Bioethics, *Human Tissue: Ethical and Legal Issues* (Nuffield Council on Bioethics London 1995) [9.9]. For a similar view, see: K Stern, 'Strict Liability and the Supply of Donated Gametes' [1994] *Medical Law Review* 261, 264.

legislation was drafted on the basis of a 'property-like structure' for the possession of parts and products of the human body.[186] While the HFEA does provide extensive control rights that are akin to property rights, the statute does not create explicit property rights. In this way, the HFEA is a good example of the legislature's desire to provide individuals with a right to control separated bodily materials without creating property rights.

In Australian jurisdictions, semen, ova and foetal tissue are excluded from the human tissue legislation,[187] and several states have enacted specific legislation to address reproductive procedures.[188] These statutes do not expressly create property rights in respect of donated gametes. As with the English legislation, consent from gamete providers is the governing principle.[189] Property-like interests are, however, evident in parts of the legislative regimes. For example, the Western Australian legislation, through section 25 of the Human Reproductive Technology Act 1991 (WA), provides that gamete providers have rights of control 'as though personal property'. Despite this reference, the WA legislature was reluctant to classify a gamete as a normal chattel governed by the law of personal property.

There is no federal statute regulating the transfer or control of gametes in the US. Many state legislatures in the US have chosen to make artificial insemination the subject of statutory enactments in order to regulate gamete donation.[190] These enactments focus on removing from gamete providers all rights, obligations, or interests with respect to children born as a result of artificial insemination.[191] They neither classify donated gametes as property nor confer on gamete providers any property rights.

(b) Common Law

The legal status of gametes has not arisen for determination in English or Australian common law; it has only been considered directly in the US. American cases have utilised contract law to resolve disputes. These cases involve human reproductive materials where contracts have been entered into by the parties specifying the conditions under which their reproductive materials could be used.

Hecht v Superior Court[192] is the leading US authority. That case arose following the suicide of Hecht's boyfriend, Kane. Before taking his life, Kane had deposited

[186] P Matthews, 'The Man of Property' [1995] *Medical Law Review* 251, 255.

[187] Human Tissue Act 1983 (NSW) s 6; Transplantation and Anatomy Act 1979 (Qld) s 7; Transplantation and Anatomy Act 1983 (SA) s 7; Human Tissue Act 1985 (Tas) s 5; Human Tissue Act 1982 (Vic) s 5; Human Tissue and Transplant Act 1982 (WA) s 6; Transplantation and Anatomy Act 1978 (ACT) s 6; Human Tissue Transplant Act 1979 (NT) s 6.

[188] Infertility (Medical Procedures) Act 1984 (Vic); Reproductive Technology Act 1988 (SA); Human Reproductive Technology Act 1991 (WA).

[189] Infertility (Medical Procedures) Act 1984 (Vic) ss 12–14; Reproductive Technology Act 1988 (SA) s 10; Human Reproductive Technology Act 1991 (WA) s 26.

[190] See, for example: California Family Code § 7613 (1994); New York Domestic Relations Law § 73 (1999); Florida Statutes § 742.11 (1997).

[191] See, for example: Florida Statutes § 742.14 (1997).

[192] 20 Cal Rptr 2d 275 (Cal CA 1993).

vials of his sperm at a sperm bank in California and signed a contract which provided that, in the event of his death, the sperm bank was to release the sperm depending on the executor's wishes. Hecht was appointed as the executor of his estate. Under the will, Kane bequeathed all the rights over his stored sperm to Hecht.[193] Following Kane's death, a dispute arose between Hecht and Kane's two children over the ownership of Kane's sperm. Hecht argued that neither Kane's estate nor his children held any proprietary rights in the sperm because Kane's will provided that she was the beneficiary of the sperm.[194] The California Court of Appeals found that the sperm was properly part of Kane's estate because 'at the time of his death, the decedent had an interest, in the nature of ownership'.[195] In this way, the court recognised limited property rights in Kane's sperm by finding that Kane retained an interest in his sperm after his death. As executor, Hecht was, therefore, entitled to the sperm.

An earlier case in the US District Court raised the concept of quasi-property in relation to a frozen embryo. In *Del Zio v Columbia Presbyterian Hospital*,[196] a couple brought an action alleging their embryo had been deliberately destroyed. They sought damages for conversion of their property and the intentional infliction of emotional distress. The District Court rejected the plaintiffs' conversion claim, but upheld their claim for mental distress and made an award of US$50,000 for the loss of the embryo.[197]

Both of these authorities suggest that, while US courts have recognised rights of control over individuals who provide gametes, they have been reluctant to invoke traditional property rights. Such reluctance seems to stem from a desire not to commodify the human body.[198] Instead, US courts have employed terms such as 'in the nature of ownership' and 'quasi-property' to confer a right of control that can be protected under law.

4 Hair Clippings

Human hair is a further example of bodily material. Except for a few growing cells at the base of the root, hair is dead tissue that is composed primarily of keratin and related proteins. Even though hair samples are composed of dead cells, it is now possible to utilise hair samples to conduct DNA testing.[199] Human hair as an object has not been the subject of serious judicial consideration because in most instances, hair clippings have no real commercial value. There are exceptions.

[193] *Hecht*, ibid at 276.
[194] Ibid, at 279.
[195] Ibid, at 284.
[196] Unreported US Dist SDNY No 14 1978.
[197] Ibid, at 18–19.
[198] B Steinbock, 'Sperm as Property' in J Harris and S Holm (eds), *The Future of Human Reproduction* (Oxford, Clarendon Press, 2000) 158.
[199] D Pergament, 'It's Not Just Hair: Historical and Cultural Considerations for an Emerging Technology' (1999) 75 *Chicago–Kent Law Review* 41, 45–46.

Historian Ruth Richardson observes that hair has traditionally been sold to wig-makers.[200] Hair jewellery became a method to memorialise a deceased loved one.[201] There is also evidence that hair has been traded for centuries throughout Europe.[202] Further, there are several modern day examples. In 1970, a lock of Byron's hair was sold for £320[203] and, similarly in 1988, a lock of Admiral Lord Nelson's hair (which had been removed by his lover Lady Emma Hamilton) sold for £5,000.[204]

Early common law authority indirectly suggests that hair was recognised as property. In *Scroggan v Stewardson*,[205] a girl who was under the age of sixteen agreed to give the defendant two ounces of her hair for 6 shillings. An action in trespass was brought by the girl's father (the plaintiff). The court held that, although an infant can agree with the barber to have her hair trimmed, she cannot license its sale.[206] Judgment was therefore entered for the plaintiff. The court did not directly address the legal status of the hair. The judgment appears to proceed on the basis that human hair can legally be sold; in this case it could not because the vendor was an infant.

Subsequently, the common law addressed the legal status of hair in *R v Herbert*.[207] The defendant was convicted of stealing hair clippings, which he had cut from a female passenger in his car. The defendant argued that hair was not property. The defendant was nonetheless convicted of larceny which required a taking and carrying away. Thus, the decision characterised human hair as an object capable of being stolen. Because the defendant was convicted of larceny, it is not clear whether the magistrates thought that growing hair could be stolen or whether the hair was stolen after being cut.

DPP v Smith[208] also concerned the cutting of hair. In this case, the complainant's ex-boyfriend (Smith) had cut off hair from her head without her consent. Smith was charged with assault occasioning bodily harm contrary to section 47 of the Offences Against the Person Act 1861 UK. The Justices for Dudley acting in the Local Justice Area of Dudley found that there was no case to answer because it had not been established that the complainant had suffered actual bodily harm. On appeal by the prosecution, the question for the court was whether the cutting of the complainant's hair constituted actual bodily harm contrary to section 47 of the Offences Against the Person Act. The English High Court found that hair is part of the human body and is 'intrinsic to each individual and to the identity of

[200] R Richardson, 'Fearful Symmetry, Corpses for Anatomy: Organs for Transplantation' in SJ Youngner (ed), *Organ Transplantation: Meaning and Realities* (Madison, University of Wisconsin Press, 1996) 82.

[201] Pergament, above n 199, at 45–46.

[202] R Scott, *The Body as Property* (London, Allen Lane, 1981) 180.

[203] *The Times* (11 February 1970) 15.

[204] 'Nelson hair fetches 5,000 pounds' *The Times* (19 February 1988) 3.

[205] (1674) 3 Keb 369; 84 ER 771.

[206] *Scroggan v Stewardson* (1674) 3 Keb 369, 369; 84 ER 771, 771.

[207] [1961] JPLGR 12.

[208] [2006] EWHC 94; [2006] 1 WLR 1571.

each individual'.[209] While noting that hair above the surface of the scalp is dead tissue 'it remains part of the body and is attached to it'.[210] As a result, the court found that cutting a substantial portion of the complainant's hair was capable of constituting an assault which occasions actual bodily harm, and, therefore, ordered that the case be sent back to be heard.[211] Although the court was not required to consider the status of the hair once separated, the case at the very least demonstrates that the law recognises the significance of human hair while attached to a living person.

There is limited judicial authority in other jurisdictions. In *United States v Cox*,[212] the defendant was convicted of bank robbery. He appealed, and one of the questions for the court was whether a haircut performed in the county jail violated the defendant's Fourth Amendment right to be free from unreasonable search and seizure. The haircut was performed without a warrant, and during the haircut, the barber preserved clippings which he sent to the police for analysis. On appeal, the defendant argued that the taking of hair samples by the police constituted a search or seizure requiring a warrant. The Seventh Circuit held that, on the facts, there was no seizure that enlivened the Fourth Amendment right. Cummings CJ found that the seizure did not occur when the hair was cut, but instead at the time when the hair clippings were preserved. By this stage, Cox had voluntarily 'abandoned his property, in this case, his hair'.[213]

These authorities suggest that the common law (at least in the US) operates on the assumption that human hair, once detached from an individual, is capable of being classified as property where it is used for a specific purpose. That is, where an intention is formed and expressed that hair clippings are to be used for a particular purpose, the law accords the clippings a proprietary status. This treatment of hair clippings provides a third circumstance whereby the law may recognise property rights in relation to separated bodily materials; namely, where an intention is expressed that such materials are to be used as property. Again, Part II analyses the potential of such a principle.

5 Waste Products

Waste products, including urine, faeces, earwax and sweat, are also often thought of as superfluous matter. Usually, they are shed or cast away by individuals and are not typically thought of as having proprietary characteristics. Several common law decisions have, however, considered the legal status of these materials.

In *R v Welsh*,[214] the defendant provided a urine sample to the police, stole it back, and then poured it down the sink. Under section 9 of the Road Traffic Act

209 *Smith*, ibid at [18].
210 Ibid.
211 Ibid, at [19].
212 428 F 2d 683 (7th Cir 1970).
213 *Cox*, ibid at 688.
214 [1974] RTR 478 (CA).

1972 UK, the police retain a right to possession of a blood or urine sample. At first instance, the court viewed the urine sample as 'property' for the purpose of statutory theft provisions, and the defendant was convicted of theft of the material. On appeal as to sentence, the Court of Appeal did not question the validity of the conviction. Hence, the ruling in *Welsh* recognised property rights, albeit a statutory right to possession, over the urine sample.

There is also some US authority dealing with the legal status of waste products. In the leading case, *Venner v State of Maryland*,[215] the police seized balloons filled with hashish found in the defendant's excrement following an examination in hospital. On appeal, the Maryland Court of Appeals held that the seizure by the police was legitimate. In considering whether the defendant had abandoned his waste products, the court observed that, although an individual may assert continuing property rights in such bodily substances, it is also 'universal human custom and human experience that such things are discarded—in a legal sense, abandoned by individuals'.[216] The court declared that 'when a person does nothing and says nothing to indicate an intent to assert his right of ownership, possession, or control over such material, the only rational inference is that he intends to abandon the material'.[217] On the facts, the court held that the police were entitled to possession of the faecal matter because the defendant had not exercised or attempted to exercise any right of possession or control over it.[218] These examples of common law authority arguably recognise that waste products can be characterised as 'property', at least for certain purposes of statutory construction.

6 Summary

There is inconsistency both within and between common law jurisdictions with respect to the legal status of separated bodily materials. *Rothery* and *Roche* suggest that the general 'no property' principle—as it was developed with respect to cadavers—does not prevent the legal classification of human biological materials as property. These cases support the view that the act of detachment creates property rights. Notably, these decisions have not tested at what stage property rights are created or to whom the property rights are allocated.

In a statutory context, the HFEA appears to be based on property principles. Even though the HFEA provides for rights of control over the use and disposal of gametes, the legislature did not explicitly allocate any property rights to gamete providers.

Property rights may also be created in separated biological materials by a third circumstance implicitly recognised by the common law. The law arguably treats hair clippings as property where an intention is formed that they are to be used as

[215] 554 A 2d 483 (Md App 1976).
[216] Ibid, at 499.
[217] Ibid.
[218] Ibid.

property for a specific purpose. Rose LJ appeared to endorse a similar principle in *R v Kelly*, when he suggested that property rights may be created where a body part has a purpose beyond its mere existence.[219] The role of intention is not explicitly acknowledged in common law decisions and commentary; it is, therefore, the subject of more detailed analysis in Part II.

D Non-proprietary Interests

Non-proprietary interests also require attention and both statutory and common law protection is considered below.

1 Statute

One potential avenue for non-proprietary protection of separated biological materials arises under the Data Protection Act 1998 UK (DPA), which provides statutory protection of personal information. The DPA replaced the Data Protection Act 1984 UK and implemented the European Directive on the processing of personal data (Personal Data Directive).[220] The DPA establishes a statutory scheme to regulate the processing of personal data that relate to living, identifiable individuals in England.[221] Due to the complexity of the legislation, litigants have been reluctant to invoke the DPA. Instead they have based their claims, wherever possible, on common law actions, particularly the duty of confidence.[222] As a result, there is limited judicial consideration of the DPA.

In outline, there are several provisions of the DPA which are relevant to the issue of non-proprietary interests in separated biological materials. First, data controllers (including private and public sector bodies and the government) who process personal data must comply with the DPA. The DPA gives individual data subjects a number of rights. These include: rights to access personal data held by data controllers;[223] limited rights to prevent processing likely to cause substantial unwarranted damage or distress;[224] a right to seek the correction or destruction of inaccurate information;[225] and a right to compensation where the DPA is violated.[226] The DPA also sets out eight Data Protection Principles. The first of these

[219] [1999] QB 621, 631.

[220] Directive 95/46/EC of the European Parliament and of the Council on the Protection of Individuals with Regard to the Processing of Personal Data and on the Free Movement of such Data [1995] OJ L/281, 31–50.

[221] *R v Brown* [1996] AC 543 (HL) 557 (Lord Hoffmann) (decided under the 1984 Act).

[222] W Lowrance, *Learning from Experience: Privacy and the Secondary Use of Data in Health Research* (London, Nuffield Trust, 2002) 17.

[223] DPA, ss 7, 8.

[224] DPA, s 10.

[225] DPA, ss 12A, 14.

[226] DPA, s 13.

seeks to ensure the fair and proportionate use of personal data. All processing of personal data must comply with these principles.[227]

Protection under the DPA raises the difficult question of defining the distinction between biological materials and personal genetic information extracted from them. In the context of data protection, the key question is whether biological materials *per se* should be classed as 'personal data' under the DPA. It is not clear under the EU Personal Data Directive whether biological materials should be treated as personal data, for the definition of 'personal data' has been left to individual countries.[228] Denmark is the only EU country to define tissue samples as personal data under data protection legislation.[229]

In order for biological materials (or extracted DNA) to be 'personal data' under the DPA, the materials must be classified as both data and personal. The DPA defines 'data' expansively to include all information recorded, held or otherwise processed automatically or held on a relevant filing system.[230] The question is whether biological materials themselves can be classified as information. The better view is that they are not information because biological materials cannot be classified as data. In *R (S) v Chief Constable of South Yorkshire Police*, Baroness Hale noted (*obiter*) that 'it could be said that samples are not "information" '.[231] Similarly, both the Medical Research Council and the Human Genetics Commission have concluded that genetic samples are not information.[232] Widening the natural meaning of 'information' to encompass biological materials is not consistent with the natural and ordinary meaning of the term.

In Australia, the scope of the term 'information' was recently considered by the Australian Law Reform Commission (ALRC) in its Report, *Essentially Yours: The Protection of Human Genetic Information.* The ALRC considered whether genetic samples could be classified as 'personal information' under the relevant federal privacy legislation. It concluded that genetic samples were not 'information',[233] noting that as a matter of statutory interpretation, the plain and ordinary meaning of the word 'information' could not include genetic samples.[234] The Federal Government agreed and concluded that federal privacy legislation is not the

[227] DPA, ss 1, 4; sch 1.

[228] D Beyleveld, D Townend, S Rouillé-Mirza and J Wright, *The Data Protection Directive and Medical Research Across Europe* (Aldershot, Ashgate, 2005) 193.

[229] The Danish Data Protection Agency (Registerrådet) has found that biological materials should be treated as personal data under the Act on Protection of Personal Data 2000 and that a collection of biological materials is a manual register under § 1(2) and § 3(3) of that legislation: case reference 2000-321-0049. For a discussion, see: M Hartlev, 'The Implementation of Data Protection Directive 95/46/EC in Denmark' in D Beyleveld, D Townend, S Rouillé-Mirza and J Wright, *Implementation of the Data Protection Directive in Relation to Medical Research in Europe* (Aldershot, Ashgate, 2005) 60.

[230] DPA, s 1.

[231] [2004] UKHL 39; [2004] 1 WLR 2196 (HL) [70].

[232] Medical Research Council, *Human Tissue and Biological Samples for Use in Research* (Medical Research Council, London, 2001) [5.1]; Human Genetics Commission, *Inside Information* (Department of Health London 2002) [3.43].

[233] *Essentially Yours*, above n 111, at [8.2].

[234] Ibid, at [8.8].

appropriate place for regulation of genetic samples.[235] The Privacy Legislation Amendment Act 2006 (Cth) does not therefore regulate genetic samples. That Act amends the definition of health information to include genetic information about an individual in a form that is, or could be, predictive of the health of the individual or a genetic relative of the individual.[236] The Federal Government considered that the state based human tissue legislation was the more appropriate sphere for the regulation of genetic samples.[237] With the exception of New South Wales, Australian states do not include genetic samples within their statutory definitions of 'personal information'.[238] New South Wales has extended data protection to biological materials. The relevant statutory provision defines 'personal information' to include 'such things as an individual's fingerprints, retina prints, body samples or genetic characteristics'.[239] One author has suggested that this additional protection in NSW may be illusory because the operative provisions of the privacy legislation were not drafted for the purposes of biological materials.[240] For separated biological materials to be protected, the relevant legislation must specifically include biological materials within the statutory definition of information or data. In England, the DPA does not explicitly include biological materials in its definition of data. The more persuasive view is that biological materials cannot currently be classified as data for the purposes of the DPA.

The fact that biological materials do not fall within the DPA regime does not preclude the application of the DPA to genetic information extracted from biological materials. Both the processing of biological materials by extracting genetic information and the subsequent recording or holding of that information are potentially covered by the DPA. In each case, it will be necessary to determine whether the statutory conditions for the use of confidential information are established under the DPA.

2 Common Law

There is no Australian or US authority considering potential non-proprietary interests in separated biological materials. In England, a case touched on potential non-proprietary interests in the context of the Police and Criminal Evidence Act 1984 UK, which provides for the retention of fingerprints and samples. Sections 60–64 of that Act provide for the taking (and retention) of samples from anyone, whether or not charged with a criminal offence. Section 61 sets out the circumstances under which fingerprints and samples may be taken without consent.

[235] Government Response to *Essentially Yours*, above n 112, at 9.
[236] Privacy Legislation Amendment Act 2006 (Cth) s 3 and sch 2.
[237] Government Response to *Essentially Yours*, above n 112, at 9.
[238] See, for example: Information Privacy Act 2000 (Vic) s 3.
[239] Health Records and Information Privacy Act 2002 (NSW) s 5(2); Privacy and Personal Information Act 1998 (NSW) s 4(2).
[240] B Alston, 'Blood Rights: The Body and Informational Privacy' (2005) 12 *Journal of Law and Medicine* 426, 432–34.

Section 64 then deals with the retention of fingerprints and samples. Importantly, an amendment in 2001 removed the requirement that a person's fingerprints or samples must be destroyed in the event that the person is not prosecuted or is acquitted.[241] The validity of the 2001 amendment was challenged in *R (S) v Chief Constable of the South Yorkshire Police; R (Marper) v Chief Constable of the South Yorkshire Police* (*R (S)*)[242] on the basis of its alleged incompatibility with articles 8 and 14 of the ECHR.

In *R (S)*, two claimants were arrested, charged and had fingerprints and DNA samples taken from them. Both claimants were subsequently acquitted, yet the police informed the claimants that, under section 64(1A) of the Police and Criminal Evidence Act 1984, the police had the right to retain the fingerprints and samples to aid in the investigation of crime. The claimants sought judicial review of the decision on the grounds that the retention of their fingerprints and samples (when they had not been convicted of a criminal offence) was incompatible with their right to respect for their privacy under article 8 of the ECHR, and their right not to be discriminated against under article 14.

The Court of Appeal (Lord Woolf CJ, Waller and Sedley LJJ) dismissed the claim on both grounds. In relation to the article 8 claim, the court observed that, although the retention of personal material represented an invasion of the right contained in article 8(1) to respect for one's private life, the purpose of retention (the prevention of crime) justified such retention under article 8(2).[243] Although the case did not directly raise the legal status of the DNA samples, Lord Woolf CJ noted the significance that such samples have for individuals:

> So far as this jurisdiction is concerned, it is my view that fingerprints and DNA samples are material which is regarded as being personal to the individual from whom it is taken and so requires legal justification before it can be retained.[244]

While Lord Woolf CJ noted the potential interference with the personal interests of the individual, he did not consider the nature of this interest in any detail. The claimants' appeal was dismissed by the House of Lords, with Lord Steyn (Lord Rodger, Baroness Hale, Lord Carswell and Lord Brown agreeing) delivering the leading speech. He held that, in so far as the retention of fingerprints and DNA samples constituted an interference with article 8(1) of the ECHR, such interference could be justified under article 8(2).[245]

The observations of Baroness Hale are the most relevant for present purposes, because although she made no comment on the legal status of the DNA samples, she was particularly concerned about the potential interference with an individual's privacy that could occur through the storage of genetic samples. Baroness

[241] Criminal Justice and Police Act 2001 UK, s 82.

[242] [2002] EWCA Civ 1275; [2002] 1 WLR 3223 (*R (S)* CA); and [2004] UKHL 39; [2004] 1 WLR 2196 (*R (S)* HL).

[243] *R (S)* CA, ibid at [42], [59] and [69].

[244] Ibid, at [32].

[245] *R (S)* HL, above n 242, at [40].

Hale considered that while it was unlikely that samples themselves could be classified as information, the reason why they are stored is because they contain genetic information about their particular donor.[246] In light of this fact, Baroness Hale opined that the same privacy principles should apply to the storage and use of genetic samples as apply to personal genetic information derived from the samples.[247] Thus, according to Baroness Hale the storage of such materials also constituted an interference with an individual's privacy.

On the authority of *R (S)*, English law recognises that individuals have, at least to some extent, a continuing interest in separated biological materials. The opinions of Lord Woolf CJ and Baroness Hale suggest that individuals have personal or non-proprietary interests in such materials.

3 Summary

English law has not defined what non-proprietary interests a living person can assert over separated biological materials. It is unlikely that the DPA extends to separated biological materials. *R (S)* appears to have raised (but not explored) the possibility of non-proprietary interests. Chapter seven examines the potential development of English law to protect an individual's non-proprietary interests over separated biological materials.

E Conclusion

This chapter has developed four points. The first is that there is no common law authority in England or Australia that directly analyses whether a person can claim property rights to biological materials removed from their body. The more difficult question—namely, whether such property rights may be allocated to a living person—was only addressed by *Moore*, *Greenberg* and *Catalona*. This chapter criticised *Moore*, in which the majority decisions held that an individual could not maintain a claim for conversion. Importantly though, *Catalona* appeared to start from the assumption that the detachment of biological materials creates property rights that can be transferred by way of gift to a third party.

Secondly, both statute and common law employ property principles in relation to separated bodily materials. They do so in an *ad hoc* manner. The examination of statute and common law leads to the third observation that there are three property-creating events: (1) the work or skill exception; (2) the act of detachment; and (3) the act of detachment coupled with an intention to treat the separated biological materials as property. Part II analyses the potential of each of these circumstances to explain the creation of property rights.

[246] *R (S)* HL, above n 242, at [70].
[247] Ibid.

The final point developed in this chapter concerns non-proprietary protection. Neither the common law nor the DPA currently extends non-proprietary protection to separated biological materials. Section 34(1)(b) of the Human Tissue Act 1983 (NSW) represents statutory recognition of an individual's right to control separated biological materials. Further exploration of non-proprietary protection is reserved for Part II. Before turning to the substantive analysis in Part II, it is necessary to complete the background picture by considering the most significant development in English law: the HTA 2004.

4

Human Tissue Act 2004

The Human Tissue Act 2004 UK (HTA 2004) is the most substantial development in English law concerning the legal regulation of biological materials separated from dead bodies and living persons. A detailed overview of the statutory regime established under the HTA 2004 is not provided here.[1] Instead, the analysis focuses on the statutory recognition of property rights and non-proprietary interests. Following a brief introduction to the statutory regime, this chapter addresses four significant features of the HTA 2004: (1) the requirement of consent; (2) exceptions to the requirement of consent; (3) the prohibition on commercial dealing; and (4) the new offence relating to non-consensual DNA analysis.

A Human Tissue Act 2004

High profile reports from the Bristol Royal Infirmary and Alder Hey hospitals concerning post-mortem bodily material retention practices[2] demonstrated the unsatisfactory position in English law with respect to separated biological materials. As Brazier observed:

> There is little dissent from the view that law reform is urgent. The confused and tangled web of different statutes and outdated common law principles must be clarified, preferably in a single Act of Parliament.[3]

In response, the Government committed itself to reform and a consultation report—*Human Bodies, Human Choices*—which was released in July 2002 following a consultation process.[4] On 12 December 2003, the Government published the Human Tissue Bill 2003. Following the Bill's second reading and its review by Standing Committee G in the House of Commons, the Government published an

[1] For a general overview of the HTA 2004, see: J Zimmern, 'Consent and Autonomy in the Human Tissue Act 2004' (2007) 18 *King's Law Journal* 313; J Herring, *Medical Law and Ethics* (Oxford, OUP, 2006) ch 7; E Jackson, *Medical Law* (Oxford, OUP, 2006) 707–12, 738–44; D Price, 'The Human Tissue Act' (2005) 68 *Modern Law Review* 798; K Liddell and A Hall, 'Beyond Bristol and Alder Hey: The Future Regulation of Alder Hey' (2005) 13 *Medical Law Review* 170.

[2] See n 2 of ch 2 and accompanying text.

[3] M Brazier, *Medicine, Patients and the Law* (London, Penguin Books, 3rd edn, 2003) 479.

[4] Department of Health, *Human Bodies, Human Choices: The Law on Human Organs and Tissue in England and Wales* (Department of Health, London, 2002).

amended version on 29 June 2004. The House of Lords made further amendments, which were accepted by the House of Commons, and the HTA 2004 received Royal Assent on 15 November 2004.

The majority of the HTA 2004 provisions apply to England, Wales and Northern Ireland.[5] Scotland has enacted a similar regime through the Human Tissue (Scotland) Act 2006 asp 4.[6] The purpose of the HTA 2004 is to provide a 'consistent legislative framework for issues relating to whole body donation and the taking, storage and use of human organs and tissues'.[7] In keeping with this remit, the HTA 2004 repeals and replaces the Human Tissue Act 1961 (HTA 1961),[8] the Human Tissue Act (Northern Ireland) 1962, the Human Organ Transplants Act 1989 (Transplants Act) and the Anatomy Act 1984 (Anatomy Act). Certain matters are excluded from the HTA 2004. The Human Fertilisation and Embryology Act 1990 (HFEA) will continue to regulate gametes and embryos. The HTA 2004 does not apply to materials that are removed and retained solely for the purpose of investigating a death under the authority of a coroner.[9] Nor does the HTA 2004 address the legal requirements concerning the removal of materials from living persons. The *removal* of bodily materials continues to be governed by the common law, including the all-important principle of consent and laws relating to battery and negligence.[10]

The HTA 2004 is divided into three parts and adopts the doctrine of consent as the fundamental principle governing the lawful storage and use of relevant material.[11] 'Relevant material' is defined as any material 'which consists of or includes human cells'.[12] Part 1 establishes the consent requirements. Part 2 sets up the licensing and regulatory system, which is to be managed initially by the Human Tissue Authority and eventually by the Regulatory Authority for Tissue and Embryos (RATE).[13] The HTA 2004 requires persons to obtain a licence to conduct certain activities, including, for example, anatomical examinations and post-

[5] HTA 2004, Explanatory Notes [3].

[6] For a discussion of the Human Tissue (Scotland) Act 2006, see M Earle and NR Whitty, 'Medical Law' (2006) *Laws of Scotland, Stair Memorial Encyclopedia Reissue* [343]–[58]. The Scottish legislation responded to similar pressures in England for reform of the law in relation to the removal, retention and transplantation of human organs. The report by the Independent Review Group—*Retention of Organs at Post-Mortem in Scotland* (2001) (available at http://www.show.scot.nhs.uk, accessed on 1 June 2007)—led to the enactment of the Scottish legislation.

[7] HTA 2004, Explanatory Notes [4].

[8] 9 & 10 Eliz 2.

[9] HTA 2004, s 11.

[10] HTA 2004, Explanatory Notes [9].

[11] *Hansard* HC vol 416 col 987 (15 January 2004).

[12] HTA 2004, s 53.

[13] The creation of RATE will follow the proposed merger of the HFEA and the Human Tissue Authority under the Government's ongoing review of Arm's Length Bodies to modernise the provision of public services. A White Paper on RATE was published on 14 December 2006 and proposed the establishment of the new Regulatory Authority for Tissue and Embryos: Department of Health, *Review of the Human Fertilisation and Embryology Act* (London, HMSO, December 2006) [2.1]. The White Paper recommended that RATE combine the statutory functions of the HTA and HFEA and anticipates that RATE will be established in 2009. For a discussion, see: SMC Gibbons, 'Are UK Genetic Databases Governed Adequately? A Comparative Legal Analysis' (2007) 27 *Legal Studies* 312, 330.

mortem examinations.[14] Part 2 also brings the regulation of human organ transplants between living persons under the charge of the Human Tissue Authority and prohibits commercial dealing in 'relevant bodily material' for the purpose of transplantation. Part 3 deals with supplementary and general matters.

The Human Tissue Authority has a significant role under the HTA 2004 and was established on 1 April 2005.[15] One of the Human Tissue Authority's statutory functions is to issue codes of practice which will give practical guidance to those carrying out activities falling within the Human Tissue Authority's remit.[16] Since its establishment, the Human Tissue Authority has drafted six codes of practice and they address: (1) consent; (2) donation of organs, tissues and cells for transplantation; (3) post-mortem examination; (4) anatomical examination; (5) removal, storage and disposal of human organs and tissue; and (6) donation of allogenic bone marrow and peripheral blood stems cells for transplantation.[17] Although failure to observe any provision of a code will not be unlawful,[18] a breach may be taken into account by the Human Tissue Authority in considering the grant and revocation of licences.[19] The relevant codes of practice extend beyond consent standards and cover the removal, storage, use and destruction of relevant material, as well as the import and export of human biological materials.[20]

1 Consent Requirements

Part 1 of the HTA 2004 enshrines the doctrine of consent as the Act's guiding principle. During the Bill's second reading, Ms Winterton (the ministerial sponsor of the HTA 2004) stated in the context of organ transplantation that 'the state does not own a human body; nor do doctors or scientists. It is up to individuals to choose whether to make this important gift, for it is a gift'.[21] She observed that '[t]he underlying principle of the changes is informed consent'.[22] Other statements at the Committee stage support the central role of consent under the HTA 2004.[23] Section 1 sets out the fundamental requirement that 'appropriate consent' be obtained for specified activities. To understand the scope of section 1, it is

[14] HTA 2004, s 16.
[15] The Human Tissue Act 2004 (Commencement No 1) Order 2005, Statutory Instrument 2005 No 919, art 3.
[16] HTA 2004, s 26.
[17] These six codes of practice received parliamentary approval on 4 July 2006. See: http://www.hta.gov.uk/guidance/codes_of_practice.cfm (accessed on 1 June 2007).
[18] HTA 2004, s 28(1).
[19] HTA 2004, s 28(2).
[20] HTA 2004, s 26(2).
[21] *Hansard* HC vol 416 col 989 (15 January 2004).
[22] *Hansard* HC vol 416 col 987 (15 January 2004).
[23] See, for example: the statement by the Parliamentary Under-Secretary of State for Health, Dr Stephen Ladyman, during the Committee stage, that '[t]he principles of the Bill are that we all own our bodies, we are entitled to determine how material from our bodies is used, and we should have consented to the use made of that material': HC Standing Committee G, col 065 (27 January 2004).

necessary to define its three components: (1) 'appropriate consent'; (2) the activities that can be undertaken for particular purposes; and (3) 'relevant material'.

First, 'appropriate consent' refers to the person from whom consent must be obtained. Who that person is will vary in different circumstances. The HTA 2004 delineates a hierarchy of persons who are empowered to give consent. For a competent living adult person, 'appropriate consent' is the consent of that individual.[24] For a deceased person, consent may be provided by the deceased in writing before death, or before a post-mortem by either a person nominated by the deceased for that purpose[25] or, failing that, by someone with parental responsibility (where the deceased was a child)[26] or someone who stood in a qualifying relationship with the deceased.[27]

The HTA 2004 does not detail the forms of consent required in each case. Instead, the Act charges the Human Tissue Authority with the task of providing guidelines by way of codes of practice to detail appropriate consent standards.[28] The *Code of Practice on Consent* details what information is required in relation to consent for the storage and use of tissue from a living or deceased person. The information that must be provided to a living person is extensive and includes: (1) the nature of proposed activities with respect to the tissue samples and the possible implications of the use of samples;[29] (2) whether the donor's consent is generic or limited to a specific research project;[30] and (3) whether the tissue samples will be used for research involving the commercial sector and commercial pharmaceutical companies.[31] The information that must be provided to a nominated representative or to a person in a qualifying relationship is also extensive, and

[24] HTA 2004, s 3. Or, in the case of living children who lack capacity, appropriate consent is the consent of a person having parental responsibility for them (s 2).

[25] HTA 2004, ss 3(6)(b)(2) and 4.

[26] HTA 2004, s 2(7).

[27] HTA 2004, s 3(6).

[28] HTA 2004, s 26.

[29] Human Tissue Authority, *Code of Practice on Consent* (May 2006) (available at www.hta.gov.uk/guidance/codes_of_practice.cfm, accessed on 1 June 2007) [77]:

> To give consent, patients (or the person with parental responsibility) must understand the nature and purpose of what is proposed and be able to make a balanced judgment. They should be told of any 'material' or 'significant' risks inherent in the way the sample will be obtained, how the tissue will be used and any possible implications of its use, eg, genetic tests. (If the person concerned is not a patient, and is volunteering samples purely for research, the general principles of providing adequate information should still apply).

[30] HTA, *Code of Practice on Consent* [79]:

> If identifiable tissue is to be used for research, patients should be told about any implications this may have. For example, they may be contacted by researchers, given feedback, or be asked for access to their medical records. Patients should be told whether the consent is generic (ie for use in any future research project which has ethical approval) or specific. If it is the latter, detailed information about the research project should be provided, in line with good practice.

[31] HTA, *Code of Practice on Consent* [80]:

> Patients should be told if their samples will or could be used for research involving the commercial sector. They should be given appropriate information on the range of activities and researchers which may be involved and whether these include commercial pharmaceutical companies.

includes: (1) the nature of, and reasons for, the proposed activities with respect to the tissue samples;[32] (2) the possible disclosure of information which may have implications for other family members;[33] and (3) consideration of the particular circumstances of the family.[34]

Secondly, section 1 of the HTA 2004 defines the activities that can be undertaken with appropriate consent. In summary, these are: (1) the storage and use of whole bodies and the removal, storage and use of 'relevant material' (the definition of which is discussed below) from a deceased person; and (2) the storage and use of relevant material from a living person. As noted above, Part 1 does not apply to the removal of relevant material from a living person, which will continue to be governed by the common law.[35]

The HTA 2004 also limits the authorised activities to particular purposes. These particular purposes are set out in Schedule 1 and include, for example, anatomical examination and research in connection with disorders or the functioning of the human body. The consent requirements vary as between relevant material removed during a post-mortem and relevant material provided by a living person. For relevant material removed during a post-mortem, appropriate consent is required for all Schedule 1 purposes.[36] Consent is not required, however, in the case of relevant material removed from a living person for the five purposes in Schedule 1, Part 2.[37]

The third component that establishes the scope of section 1 is the definition of 'relevant material'. Relevant material is defined in section 53 as any material 'which consists of or includes human cells'. The definition excludes gametes, embryos outside the human body and hair and nail from a living person.[38] Any material 'created outside the human body' is also excluded.[39] Unfortunately, the HTA 2004 does not

[32] HTA, *Code of Practice on Consent* [81]:

> Whether seeking consent from a nominated representative or from a person in a qualifying relationship, full and clear information should be provided to allow him/her to make a properly considered decision. This information should include the nature of the intended activities and the reasons for them. It should be borne in mind that some people will want more detail than others about, for example, post mortem procedures.

[33] HTA, *Code of Practice on Consent* [82]:

> Care should be taken regarding the possible disclosure of information, such as genetic information or HIV status, which the deceased person may not have wished to be disclosed, or which may have significant implications for other family members.

[34] HTA, *Code of Practice on Consent* [83]:

> While consent is needed from only one person in the hierarchy of qualifying relationships (paragraph 53 above), it is always important to consider the particular circumstances of the family. Information should be given to those who may need it, and discussions about the options available should involve the wider family wherever appropriate.

[35] HTA 2004, Explanatory Notes [9].

[36] These are the 12 purposes listed in Parts 1 and 2 of Sch 1: HTA 2004, s 1(a), (c).

[37] These purposes are: clinical audit; education or training relating to human health; performance assessment; public health monitoring; and quality assurance.

[38] HTA 2004, s 53(1)–(2).

[39] HTA 2004, s 54(7).

provide a definition of 'material created outside the human body' and the Explanatory Notes only mention cell-lines as one example of this sort of material.[40] As chapter one illustrated, there are many different forms of biological materials.[41] These include DNA amplified through polymerase chain reaction (PCR) technology and modified biological materials. It is not clear if the legislative intent is to include such biological materials within the definition of material created outside the human body. Unfortunately, neither the text of the HTA 2004 nor discussion in either the Explanatory Notes or parliamentary debates clarifies this issue.

Part 1 then establishes offences where persons use relevant material without consent (where consent is required under the HTA 2004), or use relevant material in a way that goes beyond the terms of the consent given. Section 5 provides that an offence is committed where appropriate consent has not been obtained. The Explanatory Notes state that 'where there is consent to use material for one purpose, it may not be used for another'.[42] Once consent is obtained, researchers may reuse the relevant material, but only so long as the new activities fall within the terms of the original consent (unless fresh consent is obtained, assuming that individuals allowed researchers to recontact them under their original consent). The breadth of the original consent given by a source becomes very important.

The creation of criminal offences is also a significant development. In particular, the section 5 offence represents an extension of the doctrine of consent. At common law, the doctrine of consent protects the bodily integrity of the individual, as chapter one outlined.[43] In that context, the doctrine of consent is concerned with the *removal* of bodily material. The new offence purports to govern what lawfully may be done with relevant material *after* it has been removed from the human body. In this way, the offence uses the doctrine of consent to provide individuals with greater protection over the subsequent use of relevant material removed from their body than that which is conferred under the common law. In Australia, section 34(1)(b) of the Human Tissue Act 1983 (NSW) represents a similar extension of the doctrine of consent.[44]

2 Exceptions to Consent Requirements

The HTA 2004 provides a number of exceptions to the 'appropriate consent' requirement. Most of these are of limited scope. There is one substantial exception that was introduced at the report stage in the Commons following pressure from the medical and scientific research community.[45] The Government amended the

[40] HTA 2004, Explanatory Notes [10].

[41] See nn 45–70 of ch 1 and accompanying text.

[42] HTA 2004, Explanatory Notes [20].

[43] See nn 112–36 of ch 1 and accompanying text.

[44] Inserted by the Human Tissue and Anatomy Legislation Amendment Act 2003 (NSW). See n 115 of ch 3 and accompanying text.

[45] C Dyer, 'Human Tissue Bill is Modified because of Research Needs' (2004) 328 *British Medical Journal* 1518, 1518.

original version of the Bill to provide an exception—now contained in section 1(8)—for research. Under section 1(8), research may be conducted without consent where it is ethically approved and the individual(s) from whom the relevant material came cannot be identified by the researcher. Under section 1(9), research falls within section 1(8) if it is: (a) ethically approved in accordance with regulations made by the Secretary of State; and (b) it is to be, or is, carried out in circumstances such that the person carrying it out is not in possession, and not likely to come into possession, of information from which the person from whose body the material has come can be identified.

The wording of section 1(8) and (9) raises two issues of note. The first concerns ethical approval. The HTA 2004 does not state how ethical approval is to be obtained. Instead, as section 1(9)(a) provides, ethical approval is governed by regulation—the Human Tissue Act 2004 (Ethical Approval, Exceptions from Licensing and Supply of Information about Transplants) Regulations (HTA Ethical Approval Regulations)—which were made on 15 April 2006 and came into force on 1 September 2006.[46] Under rule 2 of the HTA Ethical Approval Regulations, research is ethically approved for the purposes of section 1(9) where it is approved by a research ethics authority. The definition of what constitutes a research ethics authority is contained in rule 1 of the HTA Ethical Approval Regulations.[47] Importantly, research committees have no formal legal status and are not governed by statute or regulation.[48] They instead operate under Department of Health guidelines.[49] The exception thus appears to confer an expansive and largely undefined power on such bodies.

[46] Human Tissue Act 2004 (Ethical Approval, Exceptions from Licensing and Supply of Information about Transplants) Regs 2006 (SI 2006/1260).

[47] Under r 1 of the Human Tissue Act 2004 (Ethical Approval, Exceptions from Licensing and Supply of Information about Transplants) Regs 2006 (SI 2006/1260), a research ethics authority is defined as:

- (a) an ethics committee established or recognised in accordance with Part 2 of the Medicines for Human Use (Clinical Trials) Regulations 2004, or
- (b) any other committee established or person appointed—
  - (i) to advise on, or on matters which include, the ethics of research investigations on relevant material which has come from a human body, and
  - (ii) recognised for that purpose by, or on behalf of, the—
    - (aa) Secretary of State,
    - (bb) National Assembly of Wales, or
    - (cc) Department of Health, Social Services and Public Safety.

[48] Ethics committees were recognised under the Medicines for Human Use (Clinical Trials) Regs (SI 2004/1031), which commenced on 1 May 2004. This instrument provides a statutory grounding for good clinical practice, good manufacturing practice, research ethics committee review, informed consent of trial participants, and legal liability for injuries. See, in particular, Pt II of the Regs which creates the United Kingdom Ethics Committees Authority and provides for the establishment and recognition of ethics committees.

[49] For a discussion of the governance arrangements of research ethics committees, see: J McHale, 'Clinical Research' in A Grubb (ed), *Principles of Medical Law* (Oxord, OUP, 2nd edn, 2004) 858–61. For a discussion of the operation of research ethics committees in practice, see: SMC Gibbons et al, 'Governing Genetic Databases: Challenges Facing Research Regulation and Practice' (2007) 34 *Journal of Law and Society* 163, 181–84.

The second interesting point in this context is that, by waiving the consent requirement in circumstances where individuals cannot be identified by researchers, the HTA 2004 offers protection similar to that which is afforded to data subjects by the Data Protection Act 1998 UK (DPA). Under the section 1(8) exception, researchers may, with ethical approval, use material without consent so long as it is not possible for them to identify the individual(s) from whom the material came. The HTA 2004 seems to presuppose that the sole concern of individuals in both contexts (namely, the use of biological materials and the processing of personal data) is to protect their personal privacy. But this assumption does not necessarily follow. Individuals may have a range of other interests beyond privacy and anonymity concerns that they wish to protect or to control in relation to separated biological materials. For example, they may not want their biological materials to be used for particular kinds of research (such as research involving animals). Although this may not be a common concern among individuals, the example demonstrates that individuals may have interests going beyond informational privacy that the introduction of the section 1(8) exception fails to recognise. The existence of these broader interests supports the need for individuals to have broader rights to control what is and is not done with (or to) materials removed from them. These broader interests are canvassed in chapter seven.

In addition, there are other more limited exceptions to the consent requirement contained in section 1. A second exception, added at the report stage in the House of Commons, appears in section 7. It provides that an application may be made to the High Court for access to relevant material without consent where such access is in the interests of individuals whose health might benefit.[50] Section 7 is intended to cover two situations. One concerns cases where individuals are not traceable but analysis of their relevant material is necessary, for example to predict the likelihood of their genetic relatives having a genetic condition.[51] Secondly, the mechanism also permits waiver of consent in exceptional circumstances where the use of relevant material from living or deceased persons is in the public interest. The House of Commons debate discussed the example where an individual dies from a new virus and consent cannot be obtained from relatives.[52] As the parliamentary debates noted, this exception is intended to be limited to a major public health crisis including highly infectious diseases such as the Ebola virus.[53] The HTA 2004 does not provide further detail.

A third, more limited, exception relates to surplus material. Section 44 of the HTA 2004 provides:

> (1) It shall be lawful for material to which subsection (2) or (3) applies to be dealt with as waste.
>
> (2) This subsection applies to any material which consists of or includes human cells and which has come from a person's body in the course of his—

[50] HTA 2004, s 7.

[51] *Hansard* HC col 26 (28 June 2004) (Ms Winterton).

[52] *Hansard* HC col 27 (28 June 2004) (Ms Winterton).

[53] Ibid.

(a) receiving medical treatment,
(b) undergoing diagnostic testing, or
(c) participating in research.

(3) This subsection applies to any relevant bodily material which—

(a) has come from the human body, and
(b) ceases to be used, or stored for use, for a purpose specified in Schedule 1.

(4) This section shall not be read as making unlawful anything which is lawful apart from this section.

The HTA 2004 does not specify what consequences flow from the categorisation of surplus materials as waste. At first impression, section 44 confers a broad power for any materials to be classified as waste in the circumstances set out in section 44(2). Earlier, chapter three explained that the common law classifies as waste any materials that have been abandoned by the individuals from whom they originated.[54] In such circumstances, courts have considered that third parties may take control of such materials and use them for other purposes, including, for example, DNA analysis.

Applying this interpretation to the section, researchers would then be entitled to claim the materials once they have been categorised as abandoned waste, and to use them for other purposes without the need for any consent. If correct, this interpretation would suggest that individuals have property rights to such materials (prior to their abandonment). However, that interpretation is inconsistent with the overall intention of the HTA 2004, which focuses on obtaining consent. A better reading suggests that, under section 44, surplus materials must be disposed of where the consent provided by the individual has not specified what is to happen to such materials. Under section 26 of the HTA 2004, the Human Tissue Authority has the power to prepare codes of practice for the 'disposal' of relevant material.[55] Although section 26 refers to disposal of 'relevant material', rather than surplus waste, it is reasonable to conclude that the Human Tissue Authority's power also relates to the disposal of waste as set out in section 44. If this were not the case, then section 44 would severely undermine the consent requirements that are central to the operation of the HTA 2004.

The fourth relevant exception provides that the HTA 2004 does not alter the coroner's power to order a post-mortem without consent of any kind.[56] Further, the HTA 2004 does not alter the legality of removing or retaining bodily materials as part of a coroner's post-mortem for the purpose of investigating the cause of death. English law does not currently regulate the retention of separated bodily materials once the coroner's jurisdiction has expired.[57] In parliamentary debate, Ms Winterton stated that the Government directed the Home Office to amend the

[54] See nn 214–18 of ch 3 and accompanying text.
[55] HTA 2004, s 26(2)(l).
[56] HTA 2004, s 10.
[57] See nn 211–18 of ch 2 and accompanying text.

Coroners Rules to require the coroner to communicate with the family about retention of bodily materials.[58] As discussed in chapter two,[59] the Coroners Rules were amended in February 2005 to make specific provision for the preservation of material removed from bodies during examinations carried out on behalf of the coroner.[60]

The fifth relevant exception does not require consent where a body or bodily material comes from a person who died 100 years or more before the HTA 2004 commenced operation.[61] A related exception covers bodies or biological materials lawfully retained under the Anatomy Act.[62]

The final exception concerns 'existing holdings'. Significantly, the HTA 2004 does not apply to the continued retention of bodies or biological materials that comprise 'existing holdings'—that is, materials that were already held by anyone prior to the HTA 2004 coming into force.[63] The Human Tissue Authority addressed 'existing holdings' in the *Code of Practice on Consent.*[64] This code provides that '[i]t is lawful to store and use for scheduled purposes, without consent, human tissue that is already held in storage for a scheduled purpose on 1 September 2006'.[65] Even though the Human Tissue Authority noted that there are no statutory consent requirements for the storage or use of tissue from existing holdings, the Authority endorsed the principle 'that research should be ethically approved by a research ethics authority and that the potential benefits must outweigh any potential harm to donors of samples'.[66]

In combination, these six exceptions demonstrate that the HTA 2004 now includes several pragmatically-driven attempts to ensure that the use of relevant material is not unduly hindered by rigid adherence to the requirement for consent. Indeed, Ms Winterton noted in closing the report stage of the debate in the House of Commons that the Bill is 'workable and pragmatic'.[67] Yet, as analysing these exceptions has highlighted, the HTA 2004 fails to settle what underlying legal principles apply in the context of biological materials removed from a living person. One central concern of the HTA 2004 is to ensure that individuals retain control over material removed from them through the doctrine of consent. However, the most substantial exception recognises that consent may be bypassed where it is presumed that there is no threat to the privacy and anonymity of an individual. Thus, the doctrine of consent does not afford complete protection to the legitimate interests of individuals.

[58] *Hansard* HC vol 423 col 423 (28 June 2004) (Ms Winterton).
[59] See n 214 of ch 2 and accompanying text.
[60] Coroners Rules 1984, rr 9, 9A, 12 and 12A.
[61] HTA 2004, s 1(5)(b).
[62] HTA 2004, s 10.
[63] HTA 2004, s 9.
[64] HTA, *Code of Practice on Consent*, above n 29.
[65] Ibid, at [112].
[66] Ibid, at [114].
[67] *Hansard* HC vol 423 col 116 (28 June 2004) (Ms Winterton).

3 Prohibition on Commercial Dealing

The prohibition on commercial dealing in 'relevant bodily material' was contained in clause 29 of the original Bill.[68] The HTA 2004 no longer utilises the phrase 'relevant bodily material' in the new clause (now section 32). This revision represented a significant policy decision. Essentially, the original clause extended the previous statutory prohibition against trafficking in transplantable organs (contained in section 1 of the Transplants Act) to cover whole bodies and all forms of 'relevant bodily material'. Following the report stage, a new section 32 was drafted limiting the prohibition against commercial dealings to 'controlled material'.[69] At the report stage, Ms Winterton explained the Government's change of position as follows:

> [I]t was clearly never our intention to interfere with commercial activities that had been lawfully and ethically carried on for many years. We therefore propose to amend the Bill to confine the offences connected with advertising and supply of human tissue for reward to transplantable tissue only.[70]

Both section 32 and the original clause 29 warrant discussion. Analysing the original clause 29 is important and illuminating, because its drafting represents the most substantial attempt to codify the work or skill exception and to widen its application.[71] The exception has been maintained in the HTA 2004 under section 32. It is not clear to what extent the exception will apply in relation to material that is intended for use in transplantation. The analysis, therefore, commences with the original clause 29.

(a) Original Clause 29

The original clause 29 would have extended the ambit of the commercial prohibition to cover all relevant bodily material. In the original version of the Bill, 'relevant bodily material' was defined in clause 29(7) and (9):

[68] The phrase 'relevant bodily material' was defined in the original Bill to include bodily material which consists of, or includes, human cells.

[69] The phrase 'controlled material' is considered in detail below but, in general terms, refers to bodily material which consists of, or includes, human cells, that is intended for use in transplantation, other than gametes, embryos and material which is the subject of property because of an application of human skill.

[70] *Hansard* HC col 115 (28 June 2004) (Ms Winterton).

[71] Some Australian states appear to have given limited recognition to the work or skill exception. See, for example: s 32(2) Human Tissue Act 1983 (NSW): '[s]ubsection (1) does not apply to or in respect of the sale or supply of tissue if the tissue has been subjected to processing or treatment and the sale or supply is made for the purposes of enabling the tissue to be used for therapeutic purposes, medical purposes or scientific purposes'. See, also: Transplantation and Anatomy Act 1983 (SA) s 35(3); Human Tissue Act 1985 (Tas) s 27(2); Transplantation and Anatomy Act 1978 (ACT) s 44(2); Human Tissue Transplant Act 1979 (NT) s 24(4).

(7) For the purposes of this section, 'relevant bodily material' means any bodily material which—

(a) consists of or includes human cells, and
(b) is not of a kind excepted under subsection (9).

. . .

(9) The following kinds of bodily material are excepted—

(a) gametes,
(b) embryos,
(c) hair and nail from the body of a living person, and
(d) material which is the subject of property because of an application of human skill.

The last exception (clause 29(9)(d)) is particularly significant here. Under clause 29, it was intended that material classified as 'property because of an application of human skill' should not be caught by the prohibition on commercial use. Clause 29(9)(d) represented a partial statutory codification of the common law work or skill exception.[72] This point was made at the Committee stage, when the Parliamentary Under-Secretary of State for Health, Dr Ladyman, explicitly stated that the Bill was not intended to amend the common law position.[73] The proposed extension was significant because it remains unsettled at common law whether the exception extends to biological materials removed from a living person. The Bill originally intended to make that extension explicit, and section 32 of the HTA 2004 also makes this extension, albeit in a more limited form. Under section 32, the exception only relates to controlled material (defined below) which is intended to be used for the purpose of transplantation. The parliamentary debates in the House of Commons did not refer to the effect that this change would have on the application of human skill. During the Second Reading debate in the House of Lords, it was suggested that section 32 still sought to codify the common law exception.[74]

No discussion of the exception is contained in the original Explanatory Notes nor features in the Second Reading debate. This is regrettable because there are inconsistencies within the common law's application of the work or skill doctrine. Perhaps more significantly, the text of both the original clause 29(9)(d) and the later version contained in section 32(9)(c) appear to go beyond what the common law authorities have established.

[72] For a detailed account of the common law work or skill exception, see nn 33–114 of ch 2 and accompanying text.

[73] HC Standing Committee G, col 215 (3 February 2004). The Parliamentary Under-Secretary for Health, Dr Ladyman, made the following observation about cl 29(9) [now s 32]: '[t]hat exception reflects the current legal position as determined by the case law, that there is no property in the human body or its parts, so they cannot be bought or sold, except where human skill has been applied. The Bill recognises, but does not change that legal position; the Bill leaves the position as it is now'.

[74] *Hansard* HL vol 664 col 388 (22 July 2004) (Lord Patel). Lord Patel noted in relation to (then) cl 32 that the 'interpretation of case law in this area is not straightforward'.

Section 32 retained the 'human skill' exception in relation to the transplantation of controlled material. The principal difference between clause 29 and section 32 is that, under section 32, the prohibition applies only in the narrower context where controlled material is intended to be used for transplantation (rather than applying across the board to all forms of relevant bodily material, and whatever the intended purpose). Section 32 establishes that property rights are created where there has been 'an application of human skill' to the controlled material. The first issue concerns what level of work or skill is required for controlled material to fall within the exception. The common law has not settled what degree of work or skill is required. Similarly, the HTA 2004 does not provide any definition of what level, kind or quality of 'human skill' is required for the exception to apply. This statutory omission is indicative of the general failure at common law (and now by statute) to identify the legal principle and standards underpinning the work or skill exception.

The lack of precision in the drafting of the exception is demonstrated by two aspects of the text of the original clause 29 that have carried over into section 32. First, the text of the HTA 2004 does not refer to the acquisition of different attributes (as required at common law). Instead, it requires that there be an 'application' of human skill to the separated material. The natural and ordinary meaning of the term 'application' is putting a thing to a use or purpose.[75] The text of the section apparently requires that skill be employed in some way. At first impression, the Act's formulation seems to go beyond the work or skill exception recognised at common law. *R v Kelly* established that at common law, property rights are created where a body part acquires different attributes by virtue of the application of skill.[76] There is no mention in the Bill's original clause 29, or in section 32, of the acquisition of different attributes as being an essential condition for the exception to apply. Such an omission is significant because under the HTA 2004, it would seem that the exception applies where there is the mere application of human skill. It may be open for the courts, consistently with the exception at common law, to interpret and apply section 32 to require, in addition, the acquisition of different attributes. Nevertheless, the potential for a broader reading of the section exists which could substantially widen the scope of the work or skill exception.

The undefined nature of the work or skill exception in the HTA 2004 also raises a secondary issue of what standard of skill should be required. It may be, for example, that allowing a drop of blood to fall onto a piece of filter paper for the purposes of analysis could be sufficient. The HTA 2004 provides no guidance. Yet the standard of skill is important and it goes to the breadth of the exception. For example, is it necessary that the person performing the skill be a qualified specialist? In *Re Organ Retention Group Litigation*, Gage J observed that the individual responsible for preserving biological materials was required to have specialist qualifications.[77]

[75] *Oxford English Dictionary* (Oxford, OUP, 2nd edn, 1989) 575.

[76] [1999] QB 621, 631.

[77] [2004] EWHC 644; [2005] QB 506, App A [11].

Furthermore, both clause 29 and section 32 require 'human skill'. It is not clear from the Explanatory Notes or parliamentary debates what is meant to be conveyed by the addition of the qualifier 'human'—which has not appeared in the case law to date—to the word skill. The insertion of the word 'human' raises the question of whether the exception is applicable in the context of work primarily performed by machines. If the original clause 29 had been retained, one potential difficulty may have arisen in the context of amplified DNA. It was not clear from the text of the Bill, Explanatory Notes or parliamentary debates whether the production of amplified DNA constituted an application of human skill. This concern is exacerbated by the fact that it is not clear at common law whether amplified DNA qualifies as sufficient work or skill. In the case of amplified DNA, the human skill that is exercised is minimal because a substantial proportion of the work is conducted by the 'thermo cycler', which is the machine used by researchers to conduct the amplification. The only substantial human activity involves the mixing of materials (the original biological material, pre-mixed solutions of DNA and water) and then placing the materials in the thermo cycler.[78] Although these concerns do not need to be addressed under section 32, they demonstrate the current uncertainty surrounding the work or skill exception.

(b) Section 32

By way of partial statutory codification, the HTA 2004 essentially reproduces the offence and penalties formerly contained in the Transplants Act relating to the removal and transplantation of organs from living donors.[79] This position is consistent with the European policy against the use of the human body or biological materials for financial gain.[80] Section 32 of the HTA 2004 does extend the offence to cover also the supply of 'controlled material' intended for transplantation purposes. Section 32(1) sets out the new prohibition as follows:

(1) A person commits an offence if he:

- (a) gives or receives a reward for the supply of, or for an offer to supply, any controlled material;
- (b) seeks to find a person willing to supply any controlled material for reward;
- (c) offers to supply any controlled material for reward;
- (d) initiates or negotiates any arrangement involving the giving of a reward for the supply of, or for an offer to supply, any controlled material;
- (e) takes part in the management or control of a body of persons corporate or unincorporated whose activities consist of or include the initiation or negotiation of such arrangements.

[78] Quiagen, *PCR Handbook* (New York, Quiagen, 2002) (available at http://www1.qiagen.com, accessed on 1 June 2007) 13.

[79] HTA 2004, ss 33, 34.

[80] European Convention on the Protection of Human Rights and Dignity of the Human Being with regard to the Application of Biology and Medicine (ETS 164, Oviedo, 4 April 1997) art 21.

The definition of 'controlled material' is important and the term is defined by section 32(8), (9) and (10):

> (8) For the purposes of this section, controlled material is any material which—
>
> (a) consists of or includes human cells;
> (b) is, or is intended to be removed, from a human body,
> (c) is intended to be used for the purpose of transplantation, and
> (d) is not of a kind excepted under subsection (9).
>
> (9) The following kinds of material are excepted—
>
> (a) gametes
> (b) embryos, and
> (c) material which is the subject of property because of an application of human skill.
>
> (10) Where the body of a deceased person is intended to be used to provide material which—
>
> (a) consists of or includes human cells, and
> (b) is not of a kind excepted under subsection (9),
>
> for use for the purpose of transplantation, the body shall be treated as controlled material for the purposes of this section.

Three sorts of human materials are excluded from the prohibition under s 32(9). The first two exceptions—gametes and embryos—are covered by the regime established by the HFEA. The final exception—material that is the subject of human skill—represents, as noted above, a statutory application of the common law work or skill exception. Cell-lines, including stem cell-lines, are also excluded from the prohibition because they are not regarded as material 'from a human body'.[81]

Section 32 also provides for three broad exceptions to this prohibition against commercial dealings in controlled material for transplantation. Their inclusion suggests that Parliament implicitly considered (although without explanation) that relevant bodily material is capable of being characterised as property over which individuals or organisations may claim property rights. The first exception permits the Human Tissue Authority to designate persons who may lawfully engage in activities otherwise prohibited under section 32(3). The exception is intended to include, for example, the National Blood Service purchasing blood from overseas.[82] The second exception excludes activities in connection with acquisitions by certain museums.[83]

[81] HTA 2004, s 54(7) and HTA 2004, Explanatory Notes [48]. For a discussion of the application of the HTA 2004 to stem cell-lines, see: Medical Research Council, *Code of Practice for the Use of Human Stem Cell Lines* (MRC, London, September 2006) (available at http://www.mrc.ac.uk, accessed on 1 June 2007) [3.1].

[82] HTA 2004, Explanatory Notes [48].

[83] HTA 2004, s 32(3)(b).

The final (and perhaps most significant) exception relates to provision for the reimbursement of expenses.[84] Section 32(6) provides that payment shall not be treated as a reward where 'it is in consideration for transporting, removing, preparing, preserving or storing controlled material'.[85] The section was not contained in the original Bill and was inserted after pressure from the research and commercial communities.[86] The sub-section allows for the possibility of commercial tissue banks by permitting licence-holders to receive more than just expenses in relation to these activities.[87] Further, section 32(7) also provides that it is not an offence to provide expenses or recompense for loss of earnings to an individual who supplies transplantable material.

The practical effect of sections 32(6) and 32(7) is to allow controlled material to be bought and sold by licence-holders under certain conditions. In this way, it is arguable that controlled material is being treated as property. Under the HTA 2004, payments cannot be made directly for the controlled material, but only for other work done, such as preparing and preserving the material. During the Commons report stage, the Minister for Health, Ms Winterton, noted that '[w]e understand that payments are not made in respect of human tissue itself'.[88] The argument presented here is that any payments made are for the necessary services provided to facilitate access to controlled material. This is an artificial distinction and is analogous to the approach of certain US courts, which have characterised the provision of blood as being a service. Payment for the provision of a service arguably masks what is in reality payment for biological materials. Indeed, the research community and biotech industry were concerned that the exception provided in section 32(7) would be inadequate, and therefore sought the addition of section 32(6).

These exceptions to the commercial prohibition demonstrate that, in certain instances, the HTA 2004 does appear (at least implicitly) to treat relevant material as property. One commentator has observed that the HTA 2004 is 'philosophically grounded in property rights and interests'.[89] However, as the inconsistencies in the HTA 2004 demonstrate, the legislature has not provided a rational structure of underlying rights. Despite setting up a purportedly comprehensive legislative scheme, the HTA 2004 does not define the legal status of either 'relevant material' or 'controlled material'. Instead, as the insertion of section 32(6) demonstrates, it has been motivated by the perceived need to provide practical solutions. While these solutions may result in short-term stability, they are not based on clearly formulated or articulated legal principles. Rather, they represent an *ad hoc* response to policy concerns and perpetuate the failure of English law to clearly define, as a matter of legal doctrine, what underlying rights exist in separated bodily materials.

[84] HTA 2004, ss 32(6), 32(7).

[85] HTA 2004, s 32(6)(a).

[86] Dyer, above n 45, at 1518; L Moss, 'Tissue Bill Amended Amid Research Fears' *Scotsman* (23 June 2004) (available at http://news.scotsman.com/latest.cfm?id=3104403, accessed on 1 June 2007).

[87] HTA 2004, Explanatory Notes [48].

[88] *Hansard* HC vol 423 col 115 (28 June 2004) (Ms Winterton).

[89] Price, above n 1, at 817.

4 Non-consensual DNA Analysis Offence

Section 45 of the HTA 2004 provides for a new criminal offence where any person holds 'bodily material' (defined below) intending that the DNA in the material be analysed without 'qualifying consent'. The offence is significant here because of the non-proprietary interests it seeks to protect. The criminal offence was originally proposed by the Human Genetics Commission,[90] and a similar offence has been recommended by the Australian Law Reform Commission (ALRC).[91] Four alternatives were considered by the ALRC in relation to criminalising activities relating to non-consensual taking of an individual's DNA: (1) obtaining bodily material without the individual's consent; (2) holding bodily material with an intention that it be used for DNA analysis without the individual's consent; (3) conducting DNA analysis on bodily material without the individual's consent; and (4) using bodily material, or information derived from it, for reward or benefit without obtaining the individual's consent. The ALRC recommended the third option,[92] which the Australian Federal Government has accepted.[93]

The HTA 2004 adopts the second option. An offence will be committed where an individual holds bodily material with an intention that the DNA in it be analysed for purposes other than those provided for in section 45. Under the definition of the offence, it is not necessary for the holder to actually perform the analysis. Holding the bodily material with the requisite intention that an analysis be conducted is sufficient.

There are two important components that determine the scope of the offence. The first concerns the materials covered by the offence. 'Bodily material' is defined as 'material which has: (a) come from the human body, and (b) consists of or includes human cells'.[94] This definition includes hair, nail and gametes as it is possible to obtain DNA from these substances. It is not clear, however, whether DNA obtained from chemical sequences generated through PCR or other amplification techniques is included in the offence. These techniques produce copies of an individual's DNA and are capable of providing the same information as the original human cell.[95]

Secondly, section 45 sets out the 'excepted purposes' for which bodily material may be held for DNA analysis without consent. These include: medical treatment or diagnosis; the functions of a coroner; preventing or detecting crime; conducting a prosecution; national security purposes; and implementing an order of a

[90] Human Genetics Commission, *Inside Information* (Department of Health, London, 2002) [3.61].

[91] Australian Law Reform Commission, *Essentially Yours: The Protection of Human Genetic Information* (ALRC 96, Australian Government Publishing Service, Canberra, 2003) [12.2].

[92] Ibid, at [12.25].

[93] Government Response to Recommendations of Australian Law Reform Commission, *Essentially Yours: The Protection of Human Genetic Information* (Commonwealth Government, Canberra, 2005) 15. The Australian Federal Government has referred the proposal for a model criminal offence relating to non-consensual genetic testing to the Standing Committee of Attorneys General (SCAG).

[94] HTA 2004, s 45(5).

[95] See nn 53–59 of ch 1 and accompanying text.

court. If material is used for any of these purposes without consent then the offence is not enlivened.[96] Certain materials also fall outside the scope of the offence. Section 45(2)(a) excludes bodily material removed from the body of a person who died before the section came into force. Section 45(2)(b) also excludes bodily material held where the person holding the material is not likely to obtain information as to the identity of the individual from whom the material came.

The interests of the individual that the HTA 2004 seeks to protect through the offence are not proprietary in nature. Significantly, the establishment of a criminal offence is not based on the unauthorised *taking* of the bodily material. Had the HTA 2004 classified bodily material as property, then any unauthorised taking of such materials would also constitute theft under the criminal law. But the HTA 2004 does not classify bodily materials as property. Instead, the unauthorised taking of bodily materials represents an invasion of an individual's bodily integrity alone, and is currently protected by the law of tort through assault and battery, and the criminal law of assault.[97] The HTA 2004 does not provide control over biological materials separated from an individual following an assault.

Consequently, it is interesting to consider what individual interests Parliament may be seeking to protect through including the DNA analysis offence. The establishment of protection effectively provides individuals with control over materials once they are removed from their bodies, and thereby enforces the law's respect for each individual's autonomy and human dignity. Specifically, the offence seeks to ensure that the privacy of individuals is not violated by non-consensual testing of bodily materials.[98] The ALRC recognised an individual's privacy interest by observing that a criminal offence would punish the 'intrusion on human dignity and autonomy'.[99] Examples of potential harm to individuals include: the use of such information by employers and insurers; unwanted disclosure of long-term health prognoses; and disruption to family relationships where parentage information is revealed. Such violations of an individual's autonomy and human dignity would, therefore, constitute an offence. Thus, Parliament has at least implicitly acknowledged that individuals retain non-proprietary interests in separated biological materials.

5 Summary

The HTA 2004 enshrines the doctrine of consent as the basis for the lawful removal, storage and use of relevant material. The doctrine of consent is used to

[96] HTA s 45(1)(a)(ii) and sch 4, Pt II.

[97] See nn 112–36 of ch 1 and accompanying text.

[98] Neither the HTA 2004 nor the Explanatory Notes contain any references to autonomy or human dignity. It is reasonable, nonetheless, to infer that the offence is intended to protect individual privacy and thereby autonomy and human dignity. The s 45 offence was drafted on the basis of a recommendation by the Human Genetics Commission (HGC). In the report *Inside Information* (Department of Health, London, 2002), the HGC specifically noted at [3.51] that the proposed offence would protect individual privacy.

[99] *Essentially Yours*, above n 91, at [12.13].

control what lawfully may be done with such materials once they are removed from deceased or living persons. Nevertheless, the HTA 2004 provides for a number of exceptions to this prohibition, including the power to reimburse individuals or organisations for the supply of materials.

The HTA 2004 is nonetheless silent as to the legal status of relevant and controlled material. This silence likely represents a statutory endorsement of the common law 'no property' principle. Implicitly, though, the HTA 2004 recognises property rights in at least two circumstances. The first is where there has been an application of human skill to controlled material intended for transplantation. Secondly, the HTA 2004 recognises that a property characterisation is also appropriate for material created outside the human body, including cell-lines. These examples demonstrate the uncertainty that remains over the underlying legal principles. As a result, the UK Parliament has missed a significant opportunity to identify and articulate the legal principles that ought to cover separated biological materials.

B Conclusion

This chapter analysed the HTA 2004 with a view to understanding how English statutory law addresses the legal status of the human body and separated bodily materials. The analysis in this chapter, and the earlier chapters in Part I, demonstrated how a combination of common law and piecemeal statutory instruments has resulted in an unclear body of jurisprudence. At the heart of this confusion lies a fundamental uncertainty as to which legal principles the law should use to regulate the human body and separated bodily materials.

Increasingly, courts are faced with questions concerning the protection of an individual's interests in separated biological materials. The law has responded in two ways. First, English law recognises that property rights are created in separated biological materials on: (1) the application of work or skill; or (2) detachment. Chapter three also noted a third circumstance, namely, that property rights are created in separated biological materials following detachment, coupled with an intention that they are to be used as property.

The second response of the law is the protection of an individual's non-proprietary interests. At common law, Baroness Hale acknowledged the legal relevance of an individual's non-proprietary interests in *R (S)*. In a statutory context, the HTA 2004 recognises that individuals maintain at least non-proprietary interests in separated biological materials through consent requirements for the use of such materials in research and through the non-consensual DNA analysis offence. Developments in Australia also support the recognition of an individual's non-proprietary interests in separated biological materials, including section 34(1)(b) of the Human Tissue Act 1983 (NSW) and the Australian Government's acceptance of the ALRC's recommendation for a non-consensual DNA analysis offence.

The development of a consistent and principled approach to the legal regulation of property rights in human biological materials is necessary to ensure the protection of individual rights and to provide certainty for institutions and companies conducting medical research with such materials. Invoking the doctrine of consent, as the HTA 2004 does, avoids the first-order legal question of whether there are circumstances that should properly be seen as creating property rights in separated biological materials. As long as this question remains unanswered, English law cannot provide the necessary foundation for a coherent structure of property rights and non-proprietary interests. Part II aims to construct such a rational structure.

Part II

5

Property Rights

This chapter poses the principal question of the book: does legal principle support the creation of property rights to biological materials on their separation from a living person or dead body? The answer involves a two-stage enquiry: (1) can physical detachment transform biological materials into things that are capable of being the subject of property rights? and (2) what circumstances support the creation of property rights to separated biological materials? In addressing the first stage of the enquiry, this chapter argues that the detachment of biological materials from a human body transforms such materials into things that are capable of being the subject of property rights. Next, the second stage of the enquiry is addressed and the chapter considers whether the work or skill exception, as it has developed in English law, can explain the creation of property rights. This book contends that the work or skill exception provides an inadequate explanation and cannot operate as a general principle for the creation of property rights in separated biological materials.

A Question

It is an open question in English law whether separated biological materials can be considered property without the application of work or skill. The US decisions of *Moore v Regents of the University of California,*[1] *Greenberg v Miami Children's Hospital Research Institute*[2] and *Washington University v Catalona*[3] fail to provide a satisfactory analysis. *Moore* is the leading US authority and the Californian Supreme Court focused primarily on balancing policy considerations as to whether individuals should be entitled to sell separated biological materials, and, consequently, overlooked the more fundamental question of *who* has the right to control separated biological materials. Similarly, commentators have not addressed this question.[4] Skene, for instance, argues that balancing policy

[1] 249 Cal Rptr 494 (Cal App 1988) and 793 P 2d 479 (Cal SC 1990).

[2] 264 F Supp 2d 1064 (US DC Florida 2003).

[3] WL 1758268 (20 June 2007) (8th Cir).

[4] See, for example: JA Bovenberg, *Property Rights in Blood, Genes and Data* (Leiden, Martinus Nijhoff Publishers, 2005) 146–47; L Skene, 'Proprietary Rights in Human Bodies, Body Parts and Tissue: Regulatory Contexts and Proposals for New Laws' (2002) 22 *Legal Studies* 102, 104.

concerns should determine whether the law should recognise individuals' rights to sell separated biological materials.[5] Further, Skene submits that the potential threat that individual property rights poses to medical research dictates against granting such rights. This approach neglects the more fundamental question of who is entitled to control separated biological materials.

The question of whether property law can determine who is entitled to such control raises two issues: (1) how are biological materials transformed into things capable of being the subject of property rights? and (2) what circumstances create property rights to biological materials? Three circumstances potentially create property rights: (1) work or skill performed on biological materials; (2) detachment of biological materials *per se*; and (3) detachment of biological materials coupled with an intention to use such materials as property.

Once these two questions are answered to determine whether property rights are created, then policy concerns as to the sale of separated biological materials should be addressed. Such policy considerations are important and should guide the content of any property rights recognised by the law. One of the standard objections to creating property rights in separated biological materials is that allowing individuals to sell such materials 'commodifies' living persons, thereby undermining their human dignity. These concerns are well-documented in academic literature[6] and have shaped the policy and legislative responses of parliaments in England, Australia and the US. In England, for example, the Human Tissue Act 2004 (HTA 2004) continues the prohibition on commercial trade in human organs (previously contained in section 1 of the Human Organ Transplants Act 1989), expanding it to include controlled material intended for use in transplantation.[7]

There are policy concerns beyond commodification-based fears. Gitter, for example, argues that the recognition of property rights would undermine the traditional notion of altruistic participation in research for the benefit of society at large.[8] These policy concerns may be valid, and even warrant limits over the content of property rights that the law confers on individuals. The law must first determine who has the right to control separated biological materials.

[5] Skene, ibid at 104.

[6] For examples of this contention, see: J Mahoney, 'The Market for Human Tissue' (2000) 86 *Virginia Law Review* 163, 164–65; SR Munzer, 'Human Dignity and Property Rights in Human Body Parts' in JW Harris (ed), *Property Problems: From Genes to Pension Funds* (London, Kluwer Law International, 1997) 25, 29; DM Wagner, 'Property Rights in the Human Body: The Commercialization of Organ Transplantation and Biotechnology' (1995) 33 *Duquesne Law Review* 931, 934; RF Chandwick, 'The Market for Bodily Parts: Kant and Duties to Oneself' (1989) *Journal of Applied Philosophy* 129, 129. For a general discussion of 'commodification', see: MJ Radin, *Contested Commodities* (Boston, Harvard University Press, 1996) 1–15; MJ Radin, 'Market-Inalienability' (1987) 100 *Harvard Law Review* 1849, 1855–59. For a critique of 'commodification' objections, see: A George, 'Marketing Humanity: Should We Allow the Sale of Human Body Parts?' (2006) 7 *UTS Law Review* 11.

[7] HTA 2004, s 32.

[8] DM Gitter, 'Ownership of Ownership Tissue: A Proposal for Federal Recognition of Human Research Participants' Property Rights in their Biological Material' (2004) 61 *Washington and Lee Law Review* 257, 323. See, also: N Duxbury, 'Do Markets Degrade?' (1996) 59 *Modern Law Review* 331, 338.

B Transformation of Biological Materials

The first issue, then, is whether separated biological materials can be classified as 'things' capable of being the subject of property rights. Such a characterisation does not necessarily mean that these 'things' *are* property. Chapter one defined 'property' as 'things' that are the subject of property rights.[9] A physical object can simply exist as a thing that is *capable* of being the subject of property rights. Indeed, this book argues that the detachment of biological materials from a human body transforms such materials into 'things' that are capable of being the subject of property rights. Both judicial authority and property theory support this proposition.

Case law has recognised that separated biological materials can be classified as 'property' for the purposes of statutory construction. In England, *R v Rothery*[10] stands as authority for the proposition that blood is 'property' capable of being stolen under section 5(1) of the Theft Act 1968.[11] Similarly, *R v Welsh*[12] held that a urine sample is 'property' for the purpose of statutory theft provisions. In Australia, *Roche v Douglas*[13] held that biological materials are property under the relevant Supreme Court Rules.[14] These authorities demonstrate that, at least in certain circumstances, courts have accepted that biological materials, once separated from individuals, are capable of being classified as 'property'. Indeed, in these cases the courts have held that the materials are not only capable of being the subject of property rights, but actually are property. The creation of property rights is a separate question and is addressed below.

Property theory also supports the contention that physical detachment transforms biological materials into 'things' capable of being the subject of property rights. This book adopts Penner's definition of a 'thing'. Penner argues that all property rights are rights to things that are only contingently connected to individuals and are not intrinsic aspects of human beings.[15] The requirement of 'separability' is central to Penner's definition of what objects are included in a definition of things. Things that are intrinsically connected to individuals, such as reputations, cannot be the subject of property rights.[16] The right to bodily integrity, as detailed in chapter one,[17] is not a property right because biological materials that are part of a living person have no existence separate from that living person. As Lord Bingham noted in *R v Bentham*,[18] '[o]ne cannot possess something which is not separate and distinct from oneself. An

[9] See nn 75–80 of ch 1 and accompanying text.
[10] [1976] RTR 550 (CA).
[11] *Rothery*, ibid at 552–53.
[12] [1974] RTR 478 (CA).
[13] (2000) 22 WAR 331 (WA SC).
[14] *Roche*, ibid at 338–39.
[15] JE Penner, *The Idea of Property* (Oxford, OUP, 1997) 111.
[16] Ibid.
[17] See nn 112–14 of ch 1 and accompanying text.
[18] [2005] UKHL 18; [2005] 1 WLR 1057.

unsevered hand or finger is part of oneself . . . A person's hand or fingers are not a thing' (emphasis added).[19]

In contrast, the physical detachment of biological materials generates the necessary normative distance between a living individual and separated biological materials. This distance enables separated biological materials to be transformed into things that are, in principle, capable of being the subject of property rights. Harris notes that, for an entity to be proprietary subject matter, a human subject must be distanced from it.[20] He argues that the act of detachment provides the necessary distancing between the human source and an organ so that a separated organ is capable, conceptually speaking, of being classified as a thing that can be the subject of property rights.[21] The physical detachment of biological materials also creates the necessary normative distance between a deceased person and separated biological materials.

Even though detachment may transform biological materials into things potentially the subject of property rights, it does not follow that property rights are necessarily created on the separation of biological materials. The creation of property rights may depend on circumstances that are different from the physical detachment. The work and skill exception, for example, creates property rights following the application of work or skill to separated biological materials. Prior to the application of work or skill, these materials are not 'property', but things that are merely capable of being the subject of property rights.

C Creation of Property Rights

The second enquiry for the creation of property rights asks: what circumstances can create property rights to separated biological materials? This chapter specifically considers whether the work or skill exception can provide a general principle to explain the creation of property rights over separated biological materials. The work or skill exception is not the only potential property-creating event and, following the analysis in this chapter, chapter six addresses two other property-creating events: (1) detachment; and (2) detachment plus intention.

To explain the work or skill exception, a hypothetical factual scenario is used as a tool for examination and analysis throughout Part II. The scenario is as follows: Biological material (X) is removed from an individual (A) by a researcher (B). B intends to use X in medical research. Before undergoing the medical procedure, A consents to the removal of X. At the same time A also consents to the use of X in medical research. B then removes X and utilises X in medical research. B subjects X to the standard preservation techniques as outlined in chapter one.[22] The

[19] *Bentham*, ibid at [8].

[20] JW Harris, *Property and Justice* (Oxford, OUP, 1996) 332.

[21] Harris, ibid at 353.

[22] See nn 47–52 of ch 1 and accompanying text.

following analysis considers how the work or skill exception can explain the creation of property rights in this factual scenario.

1 Work or Skill Exception

English courts have recognised that, by applying the work or skill exception, property rights are created in biological materials removed from dead bodies.[23] This exception has been given statutory force through the HTA 2004.[24] Unfortunately, English law has not clearly identified the underlying basis of the exception as a matter of legal principle, resulting in ongoing uncertainty as to its dimensions and application.

Chapter two highlighted the inconsistent approaches to the work or skill exception adopted in *Doodeward v Spence*,[25] *Dobson v North Tyneside Health Authority*,[26] *R v Kelly*,[27] *Leeburn v Derndorfer*[28] and *Re Organ Retention Group Litigation*.[29] For example, must there be an intention to create a novel item for the work or skill exception to apply? Or, is it sufficient that work of some sort is carried out on the biological materials? *Dobson* and *Kelly* suggested that an intention to create a novel item is required, while *Doodeward* and *Re Organ Retention* did not. Chapter four also noted that the partial statutory codification of the work or skill exception under the HTA 2004 may in some respects be broader than its current formulation at common law.[30] *Kelly*, for example, referred to body parts acquiring different attributes by virtue of the application of skill.[31] The relevant section in the HTA 2004 instead refers to 'material which is the subject of property because of an application of human skill'.[32] Section 32(9)(c) HTA 2004 does not refer to additional requirements such as the acquisition of different attributes or the creation of a new item. The mere application of human skill appears sufficient under the statutory formulation.

The inconsistencies within the common law and between the common law and the HTA 2004 arise from a failure to articulate the underlying legal doctrine. Identifying the guiding principle is necessary to ensure the logical development of the exception. The English authorities—*Dobson*, *Kelly* and *Re Organ Retention*—drew on *Doodeward*. Yet, as the analysis in chapter two demonstrated, *Doodeward* did not consider the basis of the work or skill exception in any detail.[33] Academic

[23] For a detailed discussion of the work and skill exception, see nn 33–114 of ch 2 and accompanying text.
[24] HTA 2004, s 32(9)(c).
[25] (1908) 6 CLR 406 (HCA).
[26] [1997] 1 WLR 596 (CA).
[27] [1999] QB 621.
[28] [2004] VSC 172 (Vic SC).
[29] [2004] EWHC 644; [2005] QB 506. See nn 33–114 of ch 2 and accompanying text.
[30] See nn 72–81 of ch 4 and accompanying text.
[31] *Kelly*, above n 27, at 631.
[32] HTA 2004, s 32(9)(c).
[33] See nn 33–49 of ch 2 and accompanying text.

commentary has, in general, taken the work or skill exception to be a settled doctrine in English law without exploring its foundation.[34] The following analysis considers two explanations in property law for the work or skill exception.

2 First Possession

Several authors have suggested that *Kelly* is best explained as a case of 'first possession'.[35] In *Re Organ Retention*, Gage J appeared to accept the statement in *Clerk and Lindsell on Torts*[36] that the property rights will vest in the person who alters the biological materials on the basis that they are the first possessor.[37] The first-occupancy justification postulates that the first person to occupy some previously unoccupied resource should be its owner.[38] This right stems from the Roman principle of *occupatio*, which recognises that the taking of possession of a thing that is *res nullius* represents the acquisition of ownership.[39] *Res nullius* refers to things that have never been owned or have been abandoned by their original owner.

Applying a first possession argument, once biological materials are detached they are things that, until taken into possession by someone, are not owned by anyone. The UK Nuffield Council on Bioethics and academic commentaries have drawn an analogy between separated biological materials and wild animals.[40] English law establishes that the first person to take a wild animal into possession is allocated property rights to the animal.[41] The first possessor principle is also recognised in Australia[42] and the US.[43] Taking this approach, the first person to take possession of separated biological materials is entitled to property rights in them.

[34] See, for example: JV McHale, 'Regulating Genetic Databases: Some Legal and Ethical Issues' (2004) 12 *Medical Law Review* 70, 79; D Price, 'From Cosmos and Damian to Van Velzen: The Human Tissue Saga Continues' (2003) 11 *Medical Law Review* 1, 23; K Mason and GT Laurie, 'Consent or Property? Dealing with the Body and its Parts in the Shadow of Bristol and Alder Hey' (2001) 64 *Modern Law Review* 710, 721; RS Magnusson, 'Proprietary Rights in Human Tissue' in NE Palmer and E McKendrick (eds), *Interests in Goods and Services* (London, LLP, 2nd edn, 1998) 30–31.

[35] A Tettenborn, 'Wrongful Interference with Goods' in AM Dugdale (ed), *Clerk & Lindsell on Torts* (London, Sweet & Maxwell, 19th edn, 2006) 1024; A Grubb, 'I, Me, Mine: Bodies, Parts and Property and Parts' (1998) 3 *Medical Law International* 299, 312; Nuffield Council on Bioethics, *Human Tissue: Ethical and Legal Issues* (Nuffield Council on Bioethics, London, 1995) [9.11].

[36] Tettenborn, ibid at 1024.

[37] *Re Organ Retention*, above n 29, at [156].

[38] AJ Simmons, 'Original Acquisition Justifications of Private Property' in EF Paul, FD Miller and J Paul (eds), *Property Rights* (Cambridge, CUP, 1994) 63.

[39] P Birks, 'The Roman Law Concept of Dominium and the Idea of Absolute Ownership' [1985] *Acta Juridica* 1, 8; B Nicholas, *An Introduction to Roman Law* (Oxford, Clarendon Press, 1962) 130.

[40] Nuffield Council on Bioethics, *Human Tissue: Ethical and Legal Issues*, above n 35, at [9.11]; Grubb, above n 35, at 312.

[41] W Blackstone, *Commentaries on the Laws of England* (Oxford, 9th edn, 1783) vol II, 403; *Keary v Patterson* [1939] 1 KB 471, 478 (bees); *Hamps v Darby* [1946] 2 KB 311, 320 (homing pigeons); *Young v Hichens* (1844) 6 QB 606, 607 (fish). Some animals are not capable of private ownership. See, for example: *Case of Swans* (1592) 7 Co Rep 15b, 16a; 77 ER 435, 436 (swans).

[42] *Yanner v Eaton* [1999] HCA 53; (1999) 201 CLR 351 (HCA) [25] (Gleeson CJ, Gaudron, Kirby and Hayne JJ),[80] (Gummow J).

[43] *Pierson v Post* 2 Am Dec 264 (NY SC 1805) 264–67; EC Arnold, 'Law of Possession Governing the Acquisition of Animals Ferea Naturae' (1921) 55 *American Law Review* 393, 394.

The difficulty with this analogy and explanation in the context of the work or skill exception is that, if the basis of the exception is premised on taking possession, then presumably there would be no requirement for the exercise of work or skill. Simply taking possession would be sufficient. *Kelly* clearly establishes, however, that work or skill is required under the exception.[44] This objection is sufficient to reject the proposition that the work or skill exception is an application of the 'first possession' doctrine.

3 Specification Doctrine

The doctrine of specification provides a better explanation for the work or skill exception. Specification is the principle that applies to the manufacture of new objects. Examples include the making of wine from grapes, oil from olives, or bread from wheat.[45] This chapter examines the important elements of the doctrine and then returns to consider whether specification can provide the doctrinal foundation for the work or skill exception.

(a) Background

The specification doctrine's origins stem from the Roman law classification of *specificatio*.[46] *Specificatio* is a medieval neologism from the Latin phrase *speciem facere* (to produce a new species). Roman texts refer to *speciem facere* and the principle is located in the title of Justinian's *Institutes*: 'On the acquisition of ownership'.[47]

It is important to distinguish the doctrine of specification from other legal principles that apply in the context of goods that are mixed. Mixing is concerned with the situation where identical (or at least physically compatible) goods are mixed so that it is physically impossible to separate the original goods.[48] This is a question of divisible mixtures. Examples include fluid mixtures such as oil (*confusio*) and granular mixtures such as wheat (*commixitio*).[49] The principal question in the context of mixtures is defining what interest individuals have in a mixed mass to which they contributed constituent elements in ascertained proportions.

A second important distinction lies in separating the doctrine of specification from that of accession. There is sometimes an overlap between the two principles because both are original modes of acquisition. Accession involves the bringing together of entities that belong to different people with the result that one contribution merges into the other and loses its legal identity.[50] In English law, accession

[44] *Kelly*, above n 27, at 631.

[45] Blackstone, above n 41, at vol II, 404.

[46] Nicholas, above n 39, at 136; JH Baker, *An Introduction to English Legal History* (London, Butterworths, 4th edn, 2002) 381.

[47] Justinian, *Institutes*, 2.1.25.

[48] M Bridge, *Personal Property Law* (Oxford, OUP, 3rd edn, 2002) 107.

[49] *Foskett v McKeown* [2001] 1 AC 102 (HL) 121 (Lord Hope).

[50] *Rendell v Associated Finance Pty Ltd* [1957] VR 604 (Vic SC) 610; Bridge, above n 48, at 106.

is sometimes referred to as 'annexation'.[51] There are three distinct elements to accession: (1) two pieces of corporeal property are joined in some way; (2) one of the pieces (the accessory) must be subordinate to the other (the principal); and (3) attachment of the accessory to the principal must be permanent. Justinian's *Institutes* give the example of a gold thread woven into a garment.[52] In the event that one entity is attached to a principal entity, ownership of the accessory passes to the owner of the principal.[53] The principal item continues to exist in the law as a thing while the acceding item ceases to exist.

The difficult question in some cases of accession is determining which of the two items is the principal entity and formulating a test of general application is not straightforward. Academic commentary has suggested that the less valuable article should accede to the more valuable one.[54] This distinction appears inadequate, given that there may well be instances where the more valuable thing accedes to the less valuable thing. For example, if gold letters are written on paper they will accede to the paper.[55] A more satisfying test is proposed by Goode, who asserts that the principal item is the one which 'predominates as a distinct entity'.[56] For example, an engine accedes to a truck[57] and the structures built on the hull of a boat would accede to the hull.[58] Although the potential for overlap between accession and specification still exists, the principles are distinct. Accession concerns the existence of accessory and principal entities, while specification is concerned with the creation of a new thing.

No case has expressly identified specification as the principle potentially underlying the work or skill exception. Academic commentary has raised the specification principle without analysis.[59] Nevertheless, statements in *Doodeward*, *Kelly*, and *Re Organ Retention* provide, at first impression, some support for specification as the principle underlying the work or skill exception. The judgments in these cases emphasised the need for separated biological materials to change in character. In *Doodeward*, for example, Griffith CJ referred to the human body or body part as acquiring attributes that differentiate it from a human corpse awaiting burial.[60] In *Kelly*, Rose LJ also emphasised that a body part must acquire different attributes for it to be recognised as property.[61] Finally, in *Re Organ Retention*, Gage J simply noted that the work performed brought the biological

[51] *Berkley v Poulet* (1977) 241 EG 911 (CA) 913 (Scarman J).

[52] Justinian, *Institutes*, 2.1.26.

[53] *McKeown v Cavalier Yachts Pty Ltd* (1988) 13 NSWLR 303 (NSW SC) 312. See, also: P Birks, 'Mixtures' in N Palmer and E McKendrick (eds), *Interests in Goods* (London, LLP, 2nd edn, 1998) 227.

[54] E Tyler and N Palmer, *Crossley Vaines' Personal Property* (London, Butterworths, 5th edn, 1973) 432.

[55] Nicholas, above n 39, at 134.

[56] RM Goode, *Hire-Purchase Law and Practice* (London, Butterworths, 2nd edn, 1972) 751.

[57] *Lincoln Bank & Trust Co v Netter* 253 SW2d 260 (Ky CA 1952) 261.

[58] *McKeown*, above n 53, at 311–12.

[59] Mason and Laurie, above n 34, at 721.

[60] *Doodeward v Spence* (1908) 6 CLR 406 (HCA) 414.

[61] *Kelly*, above n 27, at 631.

materials within the *Doodeward* exception.[62] This requirement—that separated biological materials should acquire different attributes—is analogous to the creation of a new thing. Returning to the factual scenario presented earlier, if sufficient work or skill is performed on X, then a new thing is created which, on the basis of specification, is the subject of property rights. Accepting such a premise represents a novel application of the specification principle because, currently, the original materials (the separated biological materials) are not considered property under English law.

(b) Elements

The fundamental concern of the specification doctrine is to recognise that a new thing has been created in law. To test the strength of applying the specification doctrine to the work or skill exception, it is first necessary to consider two issues. The first is determining how and when the law should recognise that a new thing has been brought into existence. The second issue concerns the allocation of the property rights to the new thing. Once these issues are clarified, the applicability of the specification doctrine as the underlying basis for the work or skill exception can be assessed.

(i) New Thing

Specification requires that a new thing is created. The difficulty lies in determining whether there has been a sufficient change in the nature of the thing such that it can be said that a 'new thing' exists. An early decision in the US by the New York Court of Appeals held that there is 'no definite settled rule' about when the law considers a new thing to have been created.[63] There is no consideration in the Roman texts of what makes a new thing. Roman texts applied the rules of specification to five industrial processes: (1) agricultural processes (wine made from grapes); (2) smithying processes (vases made from gold); (3) tailoring (garments made from wool); (4) carpentry (cupboards made from cypresses); and (5) apothecary processes (production of perfumes).[64] In these cases, it was simply taken for granted that a new thing was created.[65] In contrast, the dyeing of wool, grapes being dried and olives being pickled were not considered by Roman texts to constitute the creation of new things.[66]

Nicholas proposes that the question of whether a new thing is created can be restated as: would the ordinary person give the thing as it is a name different from that of the thing that it was?[67] This question is too general to be of real assistance.

[62] *Re Organ Retention*, above n 29, at [257].
[63] *Silsbury v McCoon* 3 NY 379 (NY CA 1850) 390.
[64] This classification is drawn from: C van der Merwe, 'Nova Species' (2004) 2 *Roman Legal Tradition* 97, 100.
[65] JAC Thomas, 'Form and Substance in Roman Law' (1966) 19 *Current Legal Problems* 145, 151.
[66] Vinnius, *Institutes*, 2.1.25.
[67] Nicholas, above n 39, at 138.

There are difficult cases where it is not clear whether the change constitutes a mere alteration or the creation of a new thing. The paragraphs below examine English authority to discern the applicable test.

There is not a large body of jurisprudence in English law concerning the specification doctrine. Blackstone refers to the making of wine from grapes, oil from olives, and bread from wheat as examples of specification.[68] The two main English authorities on specification concern retention of title clauses. In *Borden (UK) Ltd v Scottish Timber Products Ltd*,[69] the plaintiffs supplied resin to the defendants for use in the manufacture of chipboards. Under the terms of the contract, property in the resin passed when all goods supplied to the defendants by the plaintiffs had been paid for in full. Thus, the parties considered that the resin might be used in the manufacturing process by the defendants before it had been paid for. The manufacturing process involved mixing the resin with hardeners and wax emulsion to form a 'glue mix'. The process was physically irreversible, for once the resin had been mixed with the other substances it could not be recovered. Following the appointment of a receiver and manager for the defendants, the plaintiffs brought an action to recover the sum due for the resin which they had supplied. Here, the defendants were not wrongdoers as they were entitled under the contract to use the resin before making payment. At first instance, Rubin J found that the plaintiffs were entitled to trace the resin supplied because the title had not passed and, therefore, the defendants were acting as bailee for the plaintiffs. The defendants appealed.

On appeal, the plaintiffs argued that the retention of title clause reserved to the plaintiffs the property in the resin and thereby enabled them to trace the resin through its transformations. The defendants argued that once the resin was used in the manufacturing process, it ceased to exist, and ceased to be a thing to which the retention of title clause could apply. The Court of Appeal (Buckley, Bridge and Templeman LJJ) allowed the appeal. Bridge LJ dismissed the argument that the defendants were acting as bailees for the plaintiffs and stated that the plaintiffs had no right to call for the return of the resin.[70] He accepted the contention that up to the moment when the resin was used in manufacture, the defendants held it on trust for the plaintiffs. However, title disappeared following its use, as he concluded 'there is no doubt that as soon as the resin was used in the manufacturing process it ceased to exist as resin, and accordingly the title to the resin simply disappeared'.[71]

For present purposes, the important part of Bridge LJ's statement is the assertion that title disappeared once the resin 'ceased to exist as resin'. Instead, the resin became subsumed into a new thing as one component of the chipboard. Templeman LJ also found that the resin 'ceased to exist'.[72] Bridge LJ referred to

[68] Blackstone, above n 41, at vol II, 404.
[69] [1981] Ch 25.
[70] *Borden*, ibid at 35.
[71] Ibid.
[72] Ibid, at 44.

'the production of goods of an entirely different kind',[73] and 'a wholly new substance'.[74] Buckley LJ came to a similar conclusion and his analysis gives a greater indication of why he considered there was a new thing:

> After they had so used the resin there could, in my opinion, be no property in the resin distinct from the property in the chipboard produced by the process. The manufacture had amalgamated the resin and other ingredients into a new product by an irreversible process and the resin, as resin, could not be recovered for any purpose; for all practical purposes it had ceased to exist and the ownership in that resin must also have ceased to exist.[75]

Thus, for all practical purposes, the resin had ceased to exist and could not be identified. Although the phrases used in these judgments do not offer a test of general application, they suggest that a new thing is created where the original materials cannot be restored to their original state. This position accords with the Roman law principle enunciated by Justinian. Justinian maintained that if the new thing could not be reduced to its original material, title should be allocated to the maker.[76]

The second English authority concerning specification is *Re Peachdart Ltd.*[77] There, the seller supplied leather to the buyer, a company which manufactured handbags. In the contract for sale, a title retention clause provided that, until payment was received in full, the seller retained ownership of the leather, including the right to trace any sale proceeds of any goods made with the leather. The contractual clause made the situation different from *Borden* because the relevant clause in that case did not attempt to retain title in the final product. In *Peachdart*, a receiver was appointed to the company and the receiver sought a judicial determination as to whether the title retention clause gave a priority interest to the seller in respect of the completed or uncompleted handbags. The court held that the proper construction of the contract was that the seller ceased to have exclusive title to the leather and instead had a charge over the completed and uncompleted handbags. Since the charge was unregistered, the seller had no priority.

Relevant to this discussion, the submissions of both parties focused on whether a new thing had been created by the manufacturing process. Counsel for the sellers submitted that the process of cutting and shaping of the leather into a handbag was distinguishable from the process in question in *Borden*. In *Borden*, the resin was consumed and destroyed as a separate substance when used in the manufacture of chipboard. Here, counsel argued, the items used in the manufacturing process—clasps and threads—were attached to the leather and became the sellers' property by accession. Vinelott J rejected this argument, reasoning that the accession point did not arise as he was concerned primarily with the intention of the

[73] Ibid, at 36.
[74] Ibid, at 38.
[75] Ibid, at 46.
[76] Justinian, *Institutes*, 2.1.25.
[77] [1983] 1 Ch 131.

parties.[78] In construing the contract clause, he observed that the parties could not have intended that the company would remain a bailee of each piece of leather throughout the process of manufacture. He concluded:

> The language is, I think, consistent with the view that once the process of manufacture had started so that in the course of manufacture work and materials provided by the company would result in the leather being converted into (that is incorporated in or used as material for) other goods of a distinctive character the property in those other goods would vest in [the seller] only as security for any outstanding balance of the price of the relevant parcel of leather.[79]

One possible interpretation is that Vinelott J considered the creation of goods of distinctive character sufficient to amount to the creation of a new thing. On this reading, his Honour considered that the transformation of leather into a handbag represented the creation of a new thing. For Vinelott J, the leather had ceased to exist as raw material, and the value of the handbag, thus, derived from the skill of the manufacturer. A second, and perhaps more convincing, interpretation is that Vinelott J was concerned primarily with ascertaining the intention of the parties. The statement quoted above starts by referring to the language of the contract. Consequently, the judgment is not particularly helpful in discerning what transformation is required in English law under the specification principle.

Other jurisdictions have also considered the question of whether a new thing has been created for the purposes of the specification doctrine. In Australia, *Associated Alloys Pty Ltd v Metropolitan Engineering and Fabrications Pty Ltd* considered whether steel plates that had been manufactured from steel (by a complicated manufacturing process) were new things.[80] Bryson J held that the steel plates were new things, and observed that often the question of whether a product still exists is a 'question of fact and degree'.[81]

In addition, some commentators have drawn on Scots law.[82] In general, Scots law follows Justinian's position, and *International Banking Corporation v Ferguson, Shaw & Sons*[83] provides a good demonstration of this tendency. In that case, a firm of oil merchants (Ferguson) made use of refined cottonseed oil sold to them by a party who did not have title to the oil because of a preceding security right. The oil merchants mixed the oil with their own suet and produced a new compound which was a hard compound. The Court of Session held that the compound was the property of the oil merchants because it was impossible to separate the oil from the other substances.[84] Lord Dundas set out the basic rule:

[78] *Re Peachdart*, ibid at 143.

[79] Ibid.

[80] *Associated Alloys Pty Ltd v Metropolitan Engineering and Fabrications Pty Ltd* (1996) 20 ACSR 205 (FCA) 210.

[81] Ibid, at 209.

[82] Birks, above n 53, at 228; N Palmer and A Hudson, 'Improving Stolen Chattels' in N Palmer and E McKendrick (eds), *Interests in Goods* (London, LLP, 2nd edn, 1998) 935.

[83] 1910 SC 182 (Court of Session).

[84] *International Banking Corporation*, ibid at 192.

> In this state of matters it seems to me that the case is a pure type for the application of the Roman doctrine of *specificatio*. The oil at the date of the pursuers' claim no longer existed in its original and proper form; a new species had been created, of which oil was an ingredient, but of which the component parts could not after the creation be again resolved into their original elements.[85]

This statement accords with Justinian's view and that expressed in *Borden*. A more recent decision in Scotland considered the application of specification in the context of living organisms. In *Kinloch Damph Ltd v Nordvik Salmon Farms Ltd*,[86] the court rejected an argument that the development of salmon from smolt to adult fish constituted the creation of a new thing.[87] The approach of Lord Macfadyen in this case is analysed in more detail in chapter six, which considers the application of specification to additional biological materials.[88]

The variations and complexities of manufacturing processes suggest that inevitably there will be cases where it is very difficult to determine whether a new thing has been created. Based on the analysis above, however, a new thing is created in English law where the original material ceases to exist and can no longer be restored to its original physical state. Although this is a test of general application, there will still be cases that are not easily classified.

(ii) Allocation of Property Rights

The second issue is which party should be allocated the property rights to the newly created thing in question. There are three options: (1) the owner of the original material that no longer exists; (2) the maker of the new item; and (3) no-one. In classical Roman law, there was a prominent dispute regarding who should be allocated the property rights. The Sabian school contended that the new thing belonged to the owner of the original materials while the Proculians argued that the new thing belonged to its maker.[89] Justinian upheld the Proculians' position where either the maker contributed some of the material or the character of the new thing could not be reduced to the original material.

Scots law essentially follows Justinian's rule, although there is authority that a maker cannot be the owner of the new thing where there is *male fide* manufacture.[90] This exception in Scots law is controversial; arguably, if the principle of specification precluded acquisition by a *mala fide* maker, then there would be no logical need for the specification principle. Indeed, Scottish institutional writers[91]

[85] Ibid, at 194.
[86] Unreported 30 June 1999 (Outer House Court of Session) (available at http://www.scotcourts.gov.uk/opinions/ca291499.html, accessed on 1 June 2007).
[87] *Kinloch*, ibid at 7.
[88] See nn 123–30 of ch 6 and accompanying text.
[89] Nicholas, above n 39, at 136–37.
[90] *McDonald v Provan (of Scotland Street) Ltd* (1960) SLT 231, 232. For academic support of this decision, see: T Roberts, 'A Reassessment of Historical Theories on *Specificatio* and the Requirement of Good Faith' (2002) 7 *Scottish Law and Practice Quarterly* 180.
[91] JB Nicholson (ed), J Erskine *An Institute of the Law of Scotland* (Edinburgh, 8th edn, 1871) vol II, 16.

and leading commentators[92] argue that the better view is that the property rights are allocated to the maker regardless of the maker's state of mind.

American judicial authority also supports the proposition that a wrongdoer cannot claim title to the new item in question,[93] although other US authorities disagree.[94] In *Silsbury v McCoon*,[95] the plaintiff's corn was stolen and used to make whiskey. The defendant was a creditor of the thief and seized the whiskey. Ruggles J held that the plaintiff owned the whiskey at the time it was seized,[96] thereby making the defendant liable in conversion. Ruggles J agreed that the transformation of corn into whiskey constituted the creation of a new thing. He opined that the plaintiff could only have lost ownership of the corn if the transformation had been performed by an 'honest possessor'.[97]

It is not yet settled which view English law adopts. Bracton argues that the maker should be the owner in English law.[98] The English authorities discussed above in relation to title retention indicate that the original title was lost.[99] In these cases, there was no inquiry as to wrongdoing.[100] In *Borden*, for example, Bridge LJ explicitly stated that title to the resin disappeared once the resin contributed to the formation of a new thing. In that case, title to the chipboard was allocated to the defendants.

There is one remaining English authority that touches on specification. *Glencore International v Metro Trading International*[101] concerned the question of legal title where the claimants' oil was mixed with oil in storage and blended with other types of oil to make a new product. MTI (the defendants) stored and mixed oil for sale to vessels. One set of claimants had deposited oil for storage with MTI, but when MTI became insolvent, the claimants discovered a shortfall. A bank had charges over MTI's assets and the legal question concerned whether the claimants had a proprietary claim in the oil mixture. Although *Glencore* concerned the specification doctrine, the decision focused on the allocation of property rights and not the prior question of whether a new thing had been created.

Moore-Bick J concluded that, where one person wrongfully blends his oil with oil of a different grade that belongs to another person, the new product is owned by them in common.[102] Moore-Bick J's conclusion answers the question that was

[92] GG Van der Merwe, 'Law of Things and Servitudes' in R Zimmermann and D Visser (eds), *Southern Cross: Civil Law and Common Law in South Africa* (Oxford, OUP, 1996) 724.

[93] *Silsbury*, above n 63, at 387.

[94] *Pierce v Goddard* 39 Mass 559 (Mass SC 1839) 561.

[95] 3 NY 379 (NY CA 1850).

[96] Ibid, at 390.

[97] Ibid, at 387.

[98] SE Thorne (ed), *Bracton on the Laws and Customs of England* (Cambridge, Harvard University Press, 1968) vol II, 47.

[99] *Borden*, above n 69 (resin into chipboard); *Re Peachdart*, above n 77 (leather into shapes and handbags).

[100] P Birks, 'Receipt' in P Birks and A Pretto (eds), *Breach of Trust* (Oxford, Hart Publishing, 2002) 218.

[101] [2001] 1 Lloyd's Law Reports 284 (QB).

[102] *Glencore*, ibid at [185].

left open in *Indian Oil Corporation Ltd v Greenstone Shipping SA (Panama)*:[103] namely, do the same legal principles apply to the mixing of oil where the oil of A and B is not substantially the same? Moore-Bick J concluded that the same principles should apply. As he stated:

> [W]hen one person wrongfully blends his own oil with oil of a different grade or specification belonging to another person with the result that a new product is produced, that new product is owned by them in common.[104]

Moore-Bick J noted that for this principle to apply, it would be necessary for the new thing to be fungible.[105] In the event that a *pro rata* division was not possible, Moore-Bick J considered that other principles would need to apply.[106] Moore-Bick J's analysis is novel and warrants further consideration. His conclusion appears to assert that if a wrongdoer creates a new thing, the original owner of the goods is entitled to common ownership of the new thing. There are two possible interpretations for the principle underlying his approach. The first is that Moore-Bick J purported to develop the law of specification such that the original owner of goods can still claim a title to the new thing.[107] If correct, such an approach would represent a substantial departure from English law as it is currently developing. *Glencore* concerned the mixing of different gradients of oil, arguably resulting in the creation of a new thing. A divisible mixture can amount to a specification if the mixing has created a new substance and the constituent parts are thereby destroyed. Moore-Bick J explicitly noted at several points in his judgment that a new commodity had been created.[108] But the difficulties with making the extension proposed by Moore-Bick J are demonstrated by the need to restrict his conclusion to fungible things. Moore-Bick J acknowledged that his analysis could not apply in the context of non-fungible things.[109]

A second interpretation is that his decision simply represented an application of the law of mixing, where two items of a similar nature are mixed.[110] According to this view, although a new commodity was created (through the mixing of the different gradients of oil), it was not a new thing for the purposes of the law and so the doctrine of specification was not applicable. Without a new substance, the relevant principles could legitimately be drawn from the law of tracing, which is what Moore-Bick J appeared to do.

A closer analysis of the judgment suggests that the first of these interpretations is more appropriate. Moore-Bick J noted that he was considering a novel question,[111] which was not a simple exercise in applying rules concerning mixing. He

103 [1988] 1 QB 345.

104 *Glencore*, above n 101, at [185].

105 Ibid, at [184].

106 Ibid.

107 J Ulph, 'Retaining Property Rights at Common Law Through Mixtures and Changes' [2001] *Lloyd's Commercial and Maritime Law Quarterly* 449, 453.

108 *Glencore*, above n 101, at [177], [185].

109 Ibid, at [177], [184].

110 W Swadling, 'Property' in P Birks (ed), *English Private Law* (Oxford, OUP, 2001) [4.470].

111 *Glencore*, above n 101, at [177].

considered it necessary to determine what the applicable rule should be where a divisible mixture is created from different substances. The approach he adopted—by effectively transferring a rule from the law concerning mixing to the doctrine of specification—represents a substantial departure from English law.

One difficulty with Moore-Bick J's approach is that it confuses the law of mixing with the law of specification and accession. These legal principles are intended to provide legal solutions to different questions. The law of mixing is concerned with situations where goods are substantially of the same nature and quality. The law of specification and accession concern situations where the goods are not of the same kind. In *Glencore*, Moore-Bick J purported to extend the principles established in relation to mixing to a case which should properly have been a specification case.

It is an open question whether English law will develop in the way proposed by Moore-Bick J. At present, because English law is not settled, the stronger authority remains the Court of Appeal's decision in *Borden*. The doctrine of specification permits things to be destroyed (as a matter of legal principle) even though physically the objects continue to exist. The thing as it existed in its original form is no longer recognised by the law, thus the effect of specification is that things can be destroyed unilaterally by the person who makes the new thing. In these circumstances, the original owner is left with a personal right enforceable through the tort of conversion.

Recognition of this underlying rationale for the specification principle makes it difficult to explain why the rule should be altered in circumstances where the relevant act was performed by a wrongdoer. It is possible (as a matter of logic) to say that, if a wrongdoer uses another's materials, then the rule changes and the new thing is owned by the owner of the original materials. Two problems follow from such a modification of the doctrine. The first is that applying the modification may lead to an involuntary tenancy in common on a substance that is non-fungible. Moore-Bick J identified this difficulty when he concluded that 'different principles' would need to apply in the context of a non-fungible substance.[112] Thus, Moore-Bick J's principle is not capable of general application.

The second (and more serious) problem posed by a modification for wrongdoers is that different rules will perpetuate continued legal instability. Certainty is a defining characteristic for property law principles.[113] Modification of the general principle (that the maker of the new thing is the owner at common law) in the case of wrongdoers has the potential to create uncertainty. For example, uncertainty arises for third parties who purchase a new thing in good faith. If a modification for wrongdoers is accepted, then third parties cannot not be guaranteed a secure title to the purchased product because the strength of title would vary depending on whether the maker was a conscious wrongdoer.

[112] *Glencore*, above n 101, at [184].

[113] FS Cohen, 'Dialogue on Private Property' (1954) 9 *Rutgers Law Review* 357, 366–67.

A sounder view is that, under the legal principle of specification, where a thing no longer exists at common law, property rights are allocated to the maker of the new thing. Birks argued that, although the English position is not firmly settled, *Borden* indicates that the Sabinian doctrine will not be followed.[114] Instead, it is likely that the Justinian position (granting the maker title over the new thing) will be adopted. Accordingly, following the Court of Appeal's approach in *Borden*, the maker of a new thing obtains the title to the new thing. It should be noted that the maker of the new thing may be liable under the tort of conversion if by making the new thing, they interfered with another person's property rights.[115]

(c) Application to Work or Skill Exception

With analysis of specification complete, it is now appropriate to return to the examination of the work or skill exception, and specifically the five questions posed in chapter two.[116] The first question is whether an intention to create a novel item is necessary in order for the work or skill exception to apply. Grubb, for example, suggests that the work or skill exception requires an 'intention to create a novel item with a use of its own'.[117] Under the specification doctrine, however, intention is not significant. The case law on specification illustrates that the relevant questions are twofold: whose materials were used, and who was the maker? The suggested *mala fides* rule is not currently part of English law and it is argued here that the rule should not be adopted.

The second question concerns the level of work or skill that is required to transform biological materials into property. The specification doctrine requires the creation of a new thing. Although English law is not firmly settled on this point, the analysis above suggested that a new thing is created where the original material ceases to exist and can no longer be restored to its original physical state.[118] In the context of separated biological materials, the work performed is often for preservation purposes and does not result in the creation of a new thing. In *Doodeward*, the foetus was simply preserved in a container, while in *Kelly*, although substantial work was performed on the body parts, again that work merely facilitated their preservation. Similarly, in *Re Organ Retention* the work performed to use biological materials in research was undertaken for the purpose of their preservation. Some may argue that the biological materials in that case were transformed by the preservation process and new things were thereby created. In *Re Organ Retention*, Gage J highlighted the work required by attaching to

[114] Birks, above n 53, at 228. Lawson and Rudden also suggest that '[p]robably the Roman solution would be accepted, namely, that if the object can be reduced to the original materials it would be the property of the owner of the materials, but if not, the manufacturer would become the owner': FH Lawson and B Rudden, *The Law of Property* (Oxford, Clarendon Press, 2nd edn, 1982) 75. The third edition of the *The Law of Property* (Oxford, Clarendon Press, 3rd edn, 2002) makes no specific reference to the appropriate test for the specification doctrine.

[115] Tyler and Palmer, above n 54, at 430.

[116] See nn 97–114 of ch 2 and accompanying text.

[117] Grubb, above n 35, at 250.

[118] See nn 63–88 above and accompanying text.

his judgment a summary of the process that occurs in preserving materials through the making of blocks and slides.[119] There are essentially three stages to the preservation process. First, the biological materials are embedded into a block of wax. Secondly, the block is sliced and each slice is fixed onto a slide. Thirdly, the slices are stained to show their different aspects. These steps do not cumulatively create a new entity. Even though the biological materials undergo substantial treatment, they maintain their separate physical manifestation and essential characteristics throughout the process. Indeed, the purpose of the different processes is to ensure that the biological materials are preserved in their unique state so that they can be examined microscopically. The final form in which the biological materials exist—fixed to a slide—achieves the preservation objective.

An analogy with dyeing wool supports this conclusion. The Roman jurists cited the dyeing of wool as an example of a process that did not result in the creation of a new thing.[120] Even though the tissue block creating procedure outlined in *Re Organ Retention Group Litigation*[121] is more technically advanced than the dyeing of wool, the principle remains the same. A change in form is not sufficient to create a new thing. The purpose of the specification principle is to create and allocate property rights in the event that a new thing is created. In the case of separated biological materials, entities are simply preserved rather than new ones created.

The earlier analysis of the specification principle also assists in addressing the remaining questions raised in chapter two. The third question asked: what is the nature of the property rights that the work or skill exception recognises? The authorities establish that the rights recognised are property rights and not a special or limited form of interest. The fourth question asked: should the provider of the materials or the maker of the new thing be allocated the property rights? Although this point is contentious and not firmly settled in English law, the foregoing analysis suggests that English law should maintain the current orthodoxy, allocating the property rights to the maker of the new thing. The fifth and final question from chapter two queried whether the work or skill exception is confined to materials removed from dead bodies, or also applies to materials removed from living persons. This question appears to have been answered by the HTA 2004, which partially codifies the work or skill exception in the context of controlled material removed from living persons for the purposes of transplantation.[122]

Applying the specification principle in the context of separated biological materials provides greater clarity to the ambiguities that currently exist over the work or skill exception. This analysis has demonstrated, nevertheless, that the work or skill exception is a misguided application of the specification doctrine because often the work performed is for preservation purposes and does not result in the creation of a new thing.

[119] *Re Organ Retention*, above n 29, App A.
[120] See n 66 above and accompanying text.
[121] *Re Organ Retention*, above n 29.
[122] HTA 2004, s 32(9)(c).

(d) General Principle

Two reasons militate against the notion that the work or skill exception should be employed by English law as the general principle to explain the creation of property rights to separated biological materials. First, the specification doctrine cannot be employed as the underlying principle of the work or skill exception. Secondly, the work or skill requirement has the potential to create artificial distinctions between different biological materials. Property rights are only created under the exception where the requisite degree of work or skill is performed. For example, in a research laboratory setting, the creation of property rights under the work or skill exception is dependent on the level of processing to which the biological materials have been subjected. This has the potential to create artificial distinctions between biological materials that have been subjected to different degrees of processing. In the event that some biological materials have not been subjected to a sufficient level of processing, then under the work or skill exception they cannot be accorded a proprietary status. On the other hand, biological materials that have been subjected to greater levels of processing would be accorded a proprietary status under the exception.

Notwithstanding this conclusion, the analysis of the specification and accession doctrines in this chapter remains important and informs the discussion in chapter six. There, the book contends that these doctrines can be employed to address questions concerning legal rights to additional biological materials.[123]

D Conclusion

Courts have focused primarily on policy concerns flowing from the ability of sources to sell separated biological materials when examining the question of property rights. This balancing of policy concerns should be a secondary question. Although there may be legitimate justifications for limiting what individuals are permitted to do with biological materials, the more appropriate starting point for legal analysis is to consider whether property law can determine who is entitled to control such materials. This involves a consideration of: (1) whether detachment transforms biological materials into things that are capable of being the subject of property rights; and, if so, (2) whether there are circumstances that create property rights to separated biological materials.

In respect of (1), the physical act of detachment transforms biological materials into things that are capable of being the subject of property rights. As for (2), the work or skill exception cannot provide a principle of general application to explain the creation of property rights to separated biological materials. The exception represents a misguided application of the specification doctrine. Further, because the work or skill exception is limited in application, it draws artificial distinctions.

[123] See nn 118–40 of ch 6 and accompanying text.

6

Creation and Allocation of Property Rights

A coherent and principled framework for the creation and allocation of property rights over human biological materials is important to ensure individuals retain control of materials separated from them and to provide certainty for institutions and companies using such materials in scientific and medical research. This chapter establishes a framework by examining what circumstances (other than work or skill) can support the creation of property rights. It then demonstrates how property principles can be applied in the practical context of scientific and medical research. This includes, for example, the creation and allocation of property rights over cell-lines. The analysis presented here has important implications for the regulation of biobanks, including, for example, the UK Biobank and UK Stem Cell Bank.

A Property Principles

It is first necessary to consider whether the creation of property rights can be explained by: (1) the detachment of biological materials (the 'detachment principle'); and/or (2) the detachment of biological materials coupled with an intention to use those materials as property (the 'detachment plus intention principle').

1 Detachment

The detachment of biological materials was the second property-creating event that emerged from the analysis in Part I. 'Detachment' is used here to mean the physical separation of biological materials from a human body. German law adopts detachment as a property-creating event.[1] In common law[2] and civil

[1] Bürgerliches Gesetzbuch (BGB) (German Civil Code) § 90 and Strafgesetzbuch (StGB) (German Criminal Code) § 242; HC Jaurnig (ed), *Kommentar zum StgB/BGB* (Verlag Beck München 2004) § 90/242, 33.

[2] See: C Hammond, 'Property Rights In Human Corpses and Human Tissue: The Position in Western Australia' (2002) 4 *University of Notre Dame Australia Law Review* 97, 113 (Australia); G Dworkin and I Kennedy, 'Human Tissue: Rights in the Body and Its Parts' [1993] 1 *Medical Law Review* 291, 311 (England); B Dickens, 'The Control of Living Body Materials' (1977) 27 *University of Toronto Law Journal* 142, 183 (Canada).

law[3] contexts, some academic commentary has suggested that the detachment of biological materials is a sufficient act to create property rights. This commentary has not, however, examined the act of detachment in any detail. If detachment is accepted as a property-creating event, then the detachment of biological materials satisfies both levels of enquiry outlined at the beginning of chapter five.[4] Detachment first transforms biological materials into things potentially subject to property rights; and secondly, detachment creates property rights in the separated biological materials. Applying this analysis, detachment must create property rights immediately on separation.

Chapter three noted several instances where courts have accorded separated biological materials the status of property under common law, including *R v Rothery*,[5] *R v Welsh*[6] and *Roche v Douglas*.[7] The courts in these cases were not required to consider the more difficult question of whether property rights should be allocated to the individuals from whom the materials were removed. Without making a determination on this point, these courts held that detachment created property rights to separated biological materials. On occasions when courts have considered whether property rights should be allocated to the source—*Moore v Regents of the University of California*[8] and *Greenberg v Miami Children's Hospital Research Institute*[9]—they declined to allocate these rights to the source. These decisions do not present an insurmountable hurdle to the adoption of the detachment principle as a property-creating event. Chapter three argued that these decisions did not engage in a satisfactory legal analysis.[10] Further, the most recent US decision in this area—*Washington University v Catalona*[11]—suggests that property rights in separated biological materials are initially vested in the source, and can then be transferred by way of gift to a third party.[12]

Although there is no logical necessity for the law to recognise the creation of property rights on the detachment of biological materials, three factors support the creation of such rights. The first point flows from acknowledging the protection that English law currently provides for the human body as a whole. The right to bodily integrity is the foundational principle in medical law and is protected by the law of tort.[13] The right to bodily integrity is a right enforceable generally against other members of society. It is not a property right because the law does

[3] See: NR Whitty, 'Rights of Personality, Property Rights and the Human Body in Scots Law' (2005) 9 *Edinburgh Law Review* 194, 199 (Scotland).
[4] See nn 1–8 of ch 5 and accompanying text.
[5] [1976] RTR 550 (CA).
[6] [1974] RTR 478 (CA).
[7] (2000) 22 WAR 331 (WA SC).
[8] 793 P 2d 479 (Cal SC 1990).
[9] 264 F Supp 2d 1064 (US DC Florida 2003).
[10] See nn 12–94 of ch 3 and accompanying text.
[11] WL 1758268 (20 June 2007) (8th Cir). See, also: *Washington University v Catalona* 437 F Supp 2d 985 (US DC Ed Mo 2006).
[12] See nn 68–94 of ch 3 and accompanying text.
[13] See nn 112–32 of ch 1 and accompanying text.

not classify the human body as a thing.[14] Within the rubric of property law, the human body is not considered a thing because it is viewed as an intrinsic aspect of a person.[15] An unsevered hand or finger is part of oneself and, therefore, not capable of possession.[16] Chapter five outlined a factual scenario that is employed throughout Part II.[17] In that scenario, once X is separated from A, a physical object is created that is no longer an intrinsic aspect of A. As a result, X cannot be protected under the umbrella of the right to bodily integrity. The tort of battery only protects physical invasions of the human body as a whole.

Recognising that property rights are created on detachment represents a natural extension of the right to bodily integrity. Before separation, biological materials remain part of the human body. The law provides protection for the human body, and such protection is enforceable generally against other members of society. It would seem inconsistent if the act of detachment changed biological materials from material fully protected by the law into material receiving no legal protection whatsoever.

The law can provide protection to physically detached biological materials by employing the 'bodily integrity principle'. To do this, it is necessary to create property rights. Without creating property rights, A cannot claim a right to X which is enforceable generally against other members of society. As chapter five noted, it is possible, as a matter of principle, for the law to create property rights on the detachment of biological materials, because physical detachment creates the necessary normative distancing.[18] In this way, creating property rights to separated biological materials is not dependent on the classification of the whole human body as property.

The second point supporting the creation of property rights on detachment is located in the statutory regimes governing the regulation of separated bodily materials. Statutes in England and other jurisdictions require a person's consent for the use of separated organs for transplantation.[19] In England, this statutory requirement was introduced through the Human Organ Transplants Act 1989 (Transplants Act)[20] and is now contained in the Human Tissue Act 2004 (HTA 2004).[21] These statutes recognise (at least in the transplantation context) that an individual has a right to control the use of organs once they are detached from their body. This right of control has been extended under the HTA 2004 such that an individual's consent is required (with some exceptions) for the use of relevant material in medical research.[22] The statutory consent requirement is premised on the notion that an individual has a right to control biological materials once they

[14] See nn 4–11 of ch 3 and accompanying text.
[15] JE Penner, *The Idea of Property* (Oxford, Clarendon Press, 1997) 112.
[16] *R v Bentham* [2005] UKHL 18; [2005] 1 WLR 1057, [8] (Lord Bingham).
[17] See n 22 of ch 5 and accompanying text.
[18] See nn 20–21 of ch 5 and accompanying text.
[19] See nn 97–103 of ch 3 and accompanying text.
[20] Transplants Act, s 1.
[21] HTA 2004, s 1.
[22] HTA 2004, s 1.

are detached from the human body. However, there may be cases where consent to the use of separated biological materials is not obtained. In such cases, it would be inconsistent if the law provided a right of control for consensual separation of biological materials but did not provide individuals with a right of control following non-consensual separation. To provide individuals with control following non-consensual detachment, it is necessary for property rights to be created.

The final supporting point concerns the lacuna in the law that currently exists over the legal status of separated biological materials. Without recognition of property rights, a vacuum exists in the law, whereby biological materials, once separated from a human body, are freely available to the first possessor. The law should ensure that instead of such a vacuum, a remedy exists in respect of non-consensual detachment. The creation and subsequent protection of property rights offer such a remedy.

These three points can also be employed by analogy in the context of biological materials separated from a deceased person. First, the law protects the dead body as a whole through the duty and right to bury a deceased person.[23] The right to possession of a dead body for the purposes of burial is a right enforceable generally against other members of society. As with the right to bodily integrity, the right to possession of a dead body is not a property right, because the law does not classify a dead body as a 'thing' for religious and social reasons.[24] Again, it would be inconsistent if the act of detachment changed biological materials from material protected by the law into material that receives no legal protection whatsoever. Secondly, the HTA 2004 recognises that, before death or before a post-mortem, an individual has a right to provide consent in writing as to the storage and use of the dead body.[25] The HTA 2004 also stipulates that such consent may be provided by either a person nominated by the deceased for that purpose,[26] or, failing that, by someone with parental responsibility (where the deceased was a child),[27] or someone who stood in a qualifying relationship with the deceased.[28] The HTA 2004 has recognised that these classes of person should be permitted to represent the interests of the deceased. Again, as with materials separated from living persons, it would be inconsistent if the law provided a right of control for consensual separation but did not provide representatives of the deceased with a right of control following a non-consensual separation of biological materials. Property rights are needed to provide control (and therefore a remedy at law) following non-consensual detachment.

If the act of detachment creates property rights to separated biological materials, the question remains: to whom should the property rights be allocated? An application of the detachment principle means that property rights are created

[23] See nn 161–94 of ch 2 and accompanying text.
[24] See nn 4–32 of ch 2 and accompanying text.
[25] HTA 2004, s 1.
[26] HTA 2004, ss 3(6)(b)(2), 4.
[27] HTA 2004, s 2(7).
[28] HTA 2004, s 3(6).

immediately following separation of biological materials from a living body or a deceased person. Skene argues that following detachment of biological materials, property rights should be allocated to a third party such as a researcher or hospital.[29] This book considers that, alternatively, two grounds support the allocation of property rights to the source, as the person from whom the biological materials were removed.

First, as already noted, the right to bodily integrity protects the whole body of the source. It is a right that the *source* possesses which is enforceable generally against other members of society. If the creation of property rights in separated biological materials can be viewed as an extension of the right to bodily integrity, then it follows that the right should also be allocated to the source. It would be inconsistent with the right to bodily integrity to allocate property rights over separated biological materials to any person other than the source. Allocating the property rights initially to the source acknowledges that the biological materials were originally part of that person.

Secondly, support for the allocation of the property rights to the source also exists by analogy in the English law that deals with rights over corpses and biological materials separated from corpses. Chapter two concluded that when a person dies the executor, administrator or next-of-kin has a right to custody of the corpse, which is generally enforceable against other members of society.[30] The executor can enforce this right, even though for historical reasons the law considers that a corpse is not property. English law has recognised that the executor (as the representative of the deceased) is the most appropriate person to protect the dead body. Accordingly, English law allocates to the executor the duty and corresponding right to bury the corpse.[31] This principle can be used by analogy to support allocating the property rights to the source. If the executor is allocated the right to protect a corpse (including any biological materials removed from it), then it follows that a living person should be allocated the right to control biological materials separated from their body. On the basis that property rights are created when biological materials are detached from the human body, then the source should be allocated those property rights. If the property rights were allocated to any other person, then an executor would have greater rights to biological materials removed from a corpse than a living person would have to biological materials separated from their own body. Hence, the consistent development of legal principles supports the allocation of property rights to the source.

A similar question arises in the context of biological materials that are detached from a dead body. In this context, there are two possibilities: the property right could vest in either the executor or the estate. In *Re Organ Retention*, the claimants did not submit that they had a property right to the organs that had been removed

[29] L Skene, 'Proprietary Rights in Human Bodies, Body Parts and Tissue' (2002) 22 *Legal Studies* 102, 123–27; L Skene, 'Arguments Against People Legally "Owning" Their Own Bodies, Body Parts and Tissues' (2002) 2 *Macquarie Law Journal* 165, 175.

[30] See nn 161–94 of ch 2 and accompanying text.

[31] Ibid.

from the deceased children. Instead, the claimants based their assertion of wrongful interference on the duty to bury the body.[32] *Colavito v New York Organ Donor Network Inc*[33] provides an illustration of these difficulties. There, the wife of the deceased donated both of the deceased's kidneys to the deceased's friend Peter Colavito (plaintiff). When Colavito did not receive the second kidney, he claimed, *inter alia*, that the hospital had converted his property, namely the kidney that had been gifted to him by the deceased's wife. This submission rested on the assumption that on the detachment of the kidney, the property right vested in the wife of the deceased (as part of the deceased's estate) and the wife then chose to transfer that property right by way of gift to Colavito. This assumption, that any property rights created form part of the estate, has merit. The executor essentially stands in the place of the deceased person and has several important functions. First, the executor has a limited duty to bury the deceased. This duty is limited to ensuring that the deceased person receives a dignified burial. Secondly, the executor takes charge of the deceased person's estate (including assets and liabilities) for the purpose of disposing of it according to law. The executor manages the estate on behalf of the deceased. It is therefore appropriate that any property rights created in biological materials on detachment should vest in the estate of the deceased. Any interference with such property rights can be protected by the executor.[34]

Detachment offers a clear and supportable explanation for the creation of property rights in separated biological materials. The great strength of the 'detachment principle' is the certainty it offers. That is, property rights are created and allocated to the source immediately on detachment. Nevertheless, in *Kelly*, Rose LJ suggested that English law requires more than mere detachment for property rights to be created. The potential of a third explanation—detachment coupled with intention—warrants consideration.

2 Detachment plus Intention

Property rights might also be created where biological materials are detached and a person forms and expresses an intention for those materials to be used as property. In other words, for property rights to be created, a person must form and express an intention that separated biological materials be used as property for a purpose beyond their simple existence. Other purposes could include, for example, scientific research or transplantation.

[32] *Re Organ Retention Group Litigation* [2004] EWHC 644; [2005] QB 506, [154].

[33] 438 F 3d 214 (2nd Cir 2006). See, also: *Colavito v New York Organ Donor Network* 486 F 3d 78 (2nd Cir 2007).

[34] Applying this model, the executor, therefore, takes legal title and is permitted to sue for the return of the separated biological materials. As *Commissioner of Stamp Duties (Qld) v Livingston* [1965] AC 694 emphasises, even though the executor is vested with ownership of the testator's property, the executor holds the property for the purpose of carrying out the functions and duties of the administration.

(a) Judicial Support

Two examples found in case law support the 'detachment plus intention principle'. The first relates to hair clippings that are removed from an individual. Case law in England and the US has treated hair as property.[35] Where hair is detached from a human body, its potential uses include the manufacture of wigs or DNA testing.[36] It is not surprising that there are relatively few examples of disputes over hair clippings because the commercial interests are not great.

More substantial support for a 'detachment plus intention principle' can be drawn from Rose LJ's prediction in *Kelly* that English law might develop so that a body part need not acquire different attributes before being recognised as property.[37] Rose LJ's statement has not received appropriate attention in academic literature or English case law. For Rose LJ, where separated body parts have a use beyond their mere existence, this fact may be sufficient to bring them within the law of property. If an organ has a use or significance beyond its mere existence, a reasonable inference is that the organ will be used for another purpose. A separated organ can only be used for another purpose if an intention is formed and expressed to use the organ for that purpose.

Rose LJ indicated that this was the underlying principle he had in mind when he observed that a separated body part may be considered property by English law where it is 'intended for use in an organ transplant operation, for the extraction of DNA or, for that matter, as an exhibit in a trial'.[38] This statement seems relatively uncontroversial because the examples provided are also instances in which policy considerations (such as concerns over potential commodification of the human body) are minimal. Rose LJ did not discuss, for example, the rights of an individual to sell separated organs.

(b) Intention

Detachment plus intention has the potential to provide a principle of general application for an underlying structure of property rights to separated biological materials. For property rights to be created under this principle it is necessary for a person to form and express an intention to use separated biological materials as property. In most cases, sources will form and express the requisite intention by consenting to the removal of the biological materials. The principal question then becomes whether a sufficient intention is formed and expressed. A's consent must include consent to: (a) removal of X from A's body; and (b) the construction of X as property. In most circumstances, A's consent to the removal of biological materials would also be sufficient to constitute the consent necessary for a valid expression of intent to use the materials as property. By giving consent, a source

[35] See nn 199–213 of ch 3 and accompanying text.
[36] *United States v Cox* 428 F 2d 683 (7th Cir 1970) 688.
[37] *R v Kelly* [1999] QB 621, 631.
[38] Ibid.

will usually be aware that biological materials are being removed and will be used as property for various purposes. On removal, biological materials attain a proprietary status. Intention, therefore, offers a clear explanation for the creation of property rights where a source expresses the requisite intention.

It is important here to acknowledge and reiterate the distinction between biological materials as things that are *capable* of being the subject of property rights, and things that the law regards as property. Biological materials may exist in a physical sense without being 'things' that the law regards as property. In this state, biological materials are things that can be, but are not yet, the subject of property rights. This could occur, for example, if the source does not form and express an intention prior to, or on, the detachment of biological materials. In the hypothetical factual scenario, this issue is not problematic because where A consents to the removal of X, it is likely that consent also provides the necessary intention for the law to regard X as property.

The question then arises: to whom are the property rights initially allocated? There are several options. Property rights could be allocated to the source or to the first person to take possession of the biological materials. The latter option starts from the assumption that the source does not have an automatic initial right to the property rights that are created. The source should be allocated the property rights that are created. Allocation to the source is based on the same points made above in relation to the detachment principle. First, the allocation of property rights to the source acknowledges that the biological materials came from the source and were originally protected by the right to bodily integrity. Secondly, recognition of a source's property rights is consistent with the protection that the law provides to executors with respect to dead bodies. Thirdly, the law should not allow a vacuum whereby the separated biological materials are freely available to the first possessor.

Under the 'detachment plus intention principle', it is not necessary for the biological materials to undergo a physical transformation through the application of work or skill. This precludes the danger of artificial distinctions being drawn between separated biological materials. It is acknowledged that, in property law, intention is used principally in the context of transferring property rights. However, intention can also be used to explain the creation of property rights. Rose LJ acknowledged the potential role of intention in *Kelly*.[39]

(c) Difficulties

The ability of detachment plus intention to offer a principle of general application may be limited by conceptual difficulties arising where a source does not express an intention for separated biological materials to be treated as property. These cases may be termed non-consensual separation and occur in circumstances where a source does not express an intention for separated biological materials to be used as property. Instead, another person, B, may express an intention that X

[39] Ibid.

be treated as property. If a source does not form and express the requisite intention (before or following removal of biological materials) it may be possible, at least in principle, for another person to form and express the requisite intention. In the event that a person other than the source expresses the requisite intention, it would be necessary for the other person to have possession of the separated biological materials. A requirement for possession of biological materials is necessary to ensure that the other person has the capacity to use the biological materials as property. Merely forming an intention (without possession of the materials) cannot be sufficient to create property rights in the separated biological materials.

The difficulty that then arises is whether A should still be allocated the property rights that are created even though it was B who formed and expressed an intention for the biological materials to be treated as property. Further, it may be that no-one expresses an intention to use separated biological materials as property. To assess whether this conceptual hurdle can be overcome, two examples of non-consensual separation of biological materials are considered: (1) where biological materials are removed from a source without consent; and (2) where biological materials are detached and no person forms or expresses an intention that they be used as property.

(i) Separation without Consent

If the source does not consent to the removal of biological materials, then in most cases (with the exception of an accident), a battery is committed. The tort of battery is not concerned with rights to biological materials once they are separated from a human body. Rather, battery is concerned with the protection of the human body as a whole.

The starting point for the analysis of such a scenario is to recognise that a wrong has been committed by the removal of separated biological materials without A's consent. Separation could occur, for example, if A is unconscious during a medical procedure and has not consented to the removal of biological materials. During the medical procedure, B removes biological materials. Following removal, the biological materials become things that are capable of being the subject of property rights. At the point of removal, these materials are not property. The biological materials are recognised as property as soon as a person forms and expresses an intention that they are to be used as property.

In the event that separation occurs without the consent of A, then it is likely that the wrongdoer, B, will be the person who expresses the requisite intent. Intent would be expressed usually by taking possession of the biological materials. The question then arising is: should A or B be allocated the property rights that are created? Property rights should still be allocated to A, even though B's expression of intention was the event that created the property rights.

Allocating property rights to A in these circumstances does not follow by logical necessity. The property rights must be allocated to A by default. Although there is no direct analogy in English law, support may be drawn from the conferral of property rights over wild animals. The tort of conversion does not apply to wild animals because the law classifies wild animals as *res nullius*.[40] The law recognises that the act of taking a wild animal into possession creates property rights that are allocated to the person who has possession of the wild animal. A more difficult question arises where a trespasser takes possession of a wild animal on another person's land. Under English law, property rights in the wild animal are allocated automatically to the person from whose land the animal was taken.[41] Therefore, even if a trespasser takes a wild animal into possession on another's land, the property rights are not allocated to the trespasser but to the owner of the land. The property rights are allocated to the owner of the land even though it was the trespasser who performed the act that the law recognises as the circumstance that created the property rights. In *Blades v Higgs*,[42] Lord Westbury observed that, in these circumstances, property rights must be allocated either to the owner of the land or the trespasser. He reasoned that it was unreasonable to hold that property rights should be allocated to the trespasser, concluding that they should be allocated to the owner of the soil:

> These propositions appear to me to prove clearly that game found killed by a trespasser under such circumstances as that it would be the absolute property of the owner of the soil . . . if it had been found and killed by the owner, instead of by the trespasser, does in law become the absolute property of the proprietor of the soil . . . immediately on being so caught and killed by the trespasser.[43]

Lord Westbury recognised the prior connection between the landowner and the animal. In the context of biological materials, a similar principle may be invoked by analogy. Where B removes X from A without A's consent, property rights are created over X where B forms and expresses the requisite intent. This could occur, for example, where X is removed by B while A is unconscious. Property rights to X, should not be allocated to B because the law recognises that there is a prior connection between A and X that does not exist between B and X. Accordingly, the property rights should be allocated immediately to A, even while A remains unaware of their creation. This analogy is open to criticism, because although the landowner has pre-existing property rights to the land, A has no pre-existing property rights to A's body. Property rights are only created on the detachment of biological materials from a human body. Yet A does have a prior connection to the biological materials while they remain part of A's body. This prior connection is not of a proprietary nature, but is instead recognised in, and protected by, the right to bodily integrity. This prior connection (which is analogous to that between the landowner and wild animals on the land) is arguably sufficient to support the

[40] *Pierson v Post* 2 Am Dec 264 (NY SC 1805) 264–67.

[41] *Blades v Higgs* (1865) 11 HLC 622. This common law principle was noted by the Australian High Court in *Yanner v Eaton* [1999] HCA 53; (1999) 201 CLR 351 (HCA) [25] (Gleeson CJ, Gaudron, Kirby and Hayne JJ), [80] (Gummow J).

[42] (1865) 11 HLC 622.

[43] *Blades*, ibid at 632.

allocation of property rights to the source. The same principles that were set out in relation to the 'detachment principle' apply here. In cases such as this, there are good reasons for extending the protection afforded by the bodily integrity principle.

(ii) No Intention

A second difficulty arises where biological materials are detached and no intention is formed and expressed at the time of detachment for the biological materials to be used as property. Although such circumstances may be rare, they could occur, for example, where X is detached from A following an accident. In the case of an accidental detachment, it is extremely unlikely that an intention will be formed that X is to be used as property by A or any other person.

In the case of accidental detachment, although X is transformed into a thing capable of being the subject of property rights, no property rights are created on detachment. Subsequently, the biological materials may used by a third party (C) for some other purpose. For example, C may find X and perform a DNA analysis of X. The property rights to the biological materials would be created at the point that C forms and expresses such an intention. The question arises as to whether A or C should be allocated the property rights. The property rights should be allocated to A because of A's prior connection to X. Again, this allocation of property rights does not arise by logical necessity and, as a result, property rights must be allocated to the source by default.

3 Summary

Where the source expresses the requisite intention, the 'detachment plus intention principle' offers an explanation for the creation and allocation of property rights. There are conceptual difficulties with this principle, however, where the source does not express such an intention. In light of the analysis above, this book considers that the 'detachment principle' provides the most logical and compelling explanation for the creation and allocation of property rights to separated biological materials.

B Application

The property law foundation set out in this chapter can be applied to address questions concerning biological materials and additional biological materials. The following discussion is premised on English law accepting that the 'detachment principle' creates and allocates property rights to the source. To engage in a practical analysis, it is useful to return to, and develop further, the hypothetical factual scenario first presented in chapter five.[44] In the original scenario, biological material (X) was removed from an individual (A) by a researcher (B). A consented to X's removal and authorised B to utilise X in medical research. Additional facts are as follows. In consenting to the use of X by B, A placed several conditions on the use of X. A reserved the right to demand either the return or destruction of X, and A permitted X to be used only for cancer research and for no other purposes. B accepted these conditions and proceeded to use X in research. In the course of research, B used X in four ways. First, B created amplified DNA (X_1). Secondly, B established a cell-line (X_2) using X as the original cells. Thirdly, a virus was inserted into X, which had the effect of turning X into modified biological material (X_3). B then used X_3 as the original materials to create an additional cell-line (X_4). B now wishes to pass X_1, X_2, X_3 and X_4 to a third party (C) to conduct further research into diseases other than cancer.[45]

This analysis has important practical implications because of the potential for such factual scenarios to arise in practice through the use of biological materials in projects such as the UK Biobank and UK Stem Cell Bank. Two important questions are examined. First, what legal principles govern the transfer or abandonment of property rights to original biological materials? Secondly, what legal principles apply to the creation and allocation of property rights to additional biological materials?

1 Original Biological Materials

(a) Transfer

The analysis below proceeds on the assumption that, on detachment, property rights to biological materials are created and allocated to the source. Once the property rights have been created, the source has the option to transfer the property rights in X to B by way of sale or gift. The transfer of the property rights to X cannot occur until X has been removed from A's body. Transfer of the property rights from A to B occurs immediately on removal of X. Here, an analogy may be drawn with the sale of a foal that has not yet been born. The property rights to the foal cannot pass to the purchaser until the foal has been born; but following its birth, the property rights can be transferred immediately.

This analysis also assumes that property rights created on the detachment of biological materials include the normal incidents of ownership of a chattel, such as right to transfer by sale.[46] There are significant policy concerns pertaining to the

[44] See n 22 of ch 5 and accompanying text.

[45] These different forms of biological materials were presented in ch 1. See nn 45–69 of ch 1 and accompanying text.

[46] In *Kent v Vessel 'Maria Luisa' (No 2)* [2003] FCAFC 93; (2003) 130 FCR 12 (FC FCA), Tamberlin and Hely JJ observed at [61] that: 'ordinarily the incidents of ownership of a chattel include the right to make physical use of the chattel, the right to the income from it, the power of management, and the right of alienation'.

sale of separated biological materials and such considerations may dictate that some organs should not be the subject of commercial dealings. Section 32 of the HTA 2004, for example, creates such a limitation over controlled material that is intended for transplantation. Whether this limitation is correct (as a matter of policy) is beyond the scope of this work. In the event that Parliament placed such a limitation over a source's property rights, then the property rights in X could be transferred by gift from A to a third party.

A transfer by way of gift could be achieved by obtaining the consent of the source. Indeed, a properly drafted and executed consent form could provide authority for the consensual removal of X, as well as the necessary consent to transfer the property rights. This reasoning is consistent with the finding by the Eighth Circuit Court of Appeals in *Washington University v Catalona*[47] that Washington University owned the biological materials donated to the GU Biorepository by research participants. In that case, the Eighth Circuit held that Washington University owned the biological materials on the basis of the consent forms that research participants had signed when donating the biological materials. As the court observed:

> [T]he district court properly concluded the [research participants] made informed and voluntary decisions to participate in genetic cancer research, and thereby donated their biological materials to [Washington University] as valid inter vivos gifts. This voluntary transfer of tissue and blood samples to [Washington University]—without any consideration or compensation as an incentive for doing so—demonstrates [Washington University] owns the biological samples currently housed in the Biorepository.[48]

In English law an effective gift between a donor and donee requires a clear intention by the donor to transfer the donor's rights to the donee.[49] Importantly, though, consent need not constitute a complete gift. In the factual scenario considered here, A's consent cannot constitute a complete gift. Where a source's consent limits the use of biological materials, or reserves the source's unilateral right to require that the biological materials be destroyed, then a better characterisation is that consent has created a bailment.

A bailment is a possessory relationship whereby a bailor transfers possession of a chattel to a bailee.[50] The bailment can occur at will and, in these circumstances the bailor is entitled to recall possession at any time. It is not settled under English law whether there is a separate action in bailment.[51] If there is no separate action in bailment, then other causes of action are available. At the very least, the rights

[47] WL 1758268 (20 June 2007) (8th Cir) (*Catalona* CA).

[48] Ibid, at 14 (Wollman, Riley and Shepherd JJ).

[49] *Cochrane v Moore* (1890) 25 QBD 57, 76; *Dewar v Dewar* [1975] 1 WLR 1532 (CA) 1538–39.

[50] N Palmer, 'The Place of Bailment in the Modern Law of Obligations' in N Palmer and E McKendrick (eds), *Interests in Goods* (London, LLP, 2nd edn, 1998) 461.

[51] In support of a separate action, see: Palmer, ibid at 488. For a different view, see: M Bridge, *Personal Property Law* (Oxford, OUP, 3rd edn, 2002) 33–34 and G McMeel, 'The Redundancy of Bailment' [2003] *Lloyd's Maritime and Commercial Law Quarterly* 169, 199–200.

are determined either by the terms of the contract and the proprietary relationship or, if there is no contract, by the law of tort.[52]

Returning to the factual scenario, where A reserves the right to demand the return of X, the transfer is better characterised as a bailment at will.[53] In *Washington University v Catalona*,[54] the defendants argued at first instance that by signing the consent forms to donate biological materials to Washington University, research participants made a bailment of their biological materials and not an *inter vivos* gift.[55] Limbaugh J noted that, in US law, a bailment is made on the condition that the property is to be restored to the bailor according to their directions as soon as the purpose for the bailment ceases.[56] On the facts in *Catalona*, Limbaugh J found that there was no evidence that the research participants expected the return of the biological materials.[57] On appeal, the Eighth Circuit found it unnecessary to address the bailment issue in any detail, but noted that 'the district court correctly rejected this argument as unavailing'.[58]

Two points should be noted. First, it is not clear that Limbaugh J's approach represents a correct application of the law in Missouri. The Missouri Supreme Court has held that bailments do not always require the property to be returned to the bailor.[59] Secondly, it is likely that there may be cases where the consent of the source expressly includes a right to demand the return of the biological materials. In such cases, there is no reason, at least in principle, why the law of bailment cannot apply. Thus, the form of A's consent is crucial in determining the legal consequences for a property-based model.

(b) Abandonment

A secondary question arises where no transfer of the property rights occurs at law. If there is no transfer, then A retains the property rights to X. On these facts, the only way that B (or any other recipient) could obtain the property rights to X is in the event that A abandons X. If B can establish that A has abandoned A's property rights to X, then such rights can be allocated to B. If B cannot establish that A has abandoned those rights, then B remains open to a claim in conversion.

There has been some academic discussion of the concept of abandonment in the context of biological materials.[60] The UK Nuffield Council on Bioethics suggested that biological materials removed in the course of treatment should be regarded in

[52] *Morris v CW Martin & Sons Ltd* [1966] 1 QB 716.

[53] For a discussion of bailment in the context of human tissue, see: D Nicol, 'Property in Human Tissue and the Rights of Commercialisation' (2004) 30 *Monash University Law Review* 139, 148–52.

[54] 437 F Supp 2d 985 (US DC Ed Mo 2006).

[55] *Catalona*, ibid at 1001.

[56] Ibid.

[57] Ibid.

[58] *Catalona* CA, above n 47, 14 fn 9.

[59] *State v Edwards* 137 S W 2d 447 (Mo SC 1940) 451.

[60] J McHale, 'Waste, Ownership and Bodily Products' (2000) 8 *Health Care Analysis* 123, 130–32; P Matthews, 'The Man of Property' [1995] *Medical Law Review* 251, 259–61.

law as abandoned.[61] Matthews has criticised the Nuffield Council's conclusion and argued that abandonment is 'unlikely to be implied from general consent to an operation'.[62] Furthermore, it is unclear in English law whether the doctrine of abandonment applies in relation to chattels.[63] Blackstone asserts that chattels can exist without them having an owner by noting that 'absolutely abandoned' chattels are returned to the common stock.[64] Some commentators argue that property cannot be abandoned as a matter of principle.[65] The more generally accepted view is that English law recognises a principle of abandonment.[66] Specific examples of abandonment include theft,[67] wreck[68] and salvage.[69] If it is accepted that a chattel can be abandoned, English law recognises that the first person to take possession of the abandoned chattel acquires property rights that are effective against the world (including the original owner).[70]

The difficult question in the context of abandonment is determining whether there is a clear and unequivocal intention to abandon.[71] Some US authorities have applied the abandonment doctrine in the context of materials separated from the human body. *Venner v State of Maryland*[72] concerned rights over excrement. In that case, the court held that the appellant had abandoned his excrement in accordance with the universal human practice.[73] Similarly, in *US v Cox*, the US Court of Appeals for the Seventh Circuit held that the defendant had abandoned his hair clippings by consenting to a haircut.[74]

These authorities illustrate that the law of property is capable of applying the doctrine of abandonment in the context of materials detached from a human body. Abandonment is, nonetheless, very difficult to establish as a matter of fact, and may prove to be particularly difficult in the context of medical procedures.[75] If it can be established that A has abandoned X, then the first person to take possession of X is allocated the property rights. If abandonment cannot be established, then A retains property rights to X.

61 Nuffield Council on Bioethics, *Human Tissue: Ethical and Legal Issues* (Nuffield Council on Bioethics, London, 1995) [9.14].

62 Matthews, above n 60, at 268.

63 W Swadling, 'Property' in P Birks (ed), *English Private Law* (Oxford, OUP, 2000) 383.

64 W Blackstone, *Commentaries on the Laws of England* (Oxford, 9th edn, 1783) vol II, 9, 402.

65 E Tyler and N Palmer, *Crossley Vaines' Personal Property* (London, Butterworths, 5th edn, 1973) 427.

66 *Halsbury's Laws of England* (4th edn) vol 35, [1125]; Bridge, above n 51, at 22; S Gleeson, *Personal Property Law* (London, FT Law & Tax, 1997) 51; AP Bell, *Modern Law of Personal Property in England and Ireland* (London, Butterworths, 1989) 53; A Hudson, 'Is Divesting Abandonment Possible at Common Law' (1984) 100 *Law Quarterly Review* 110, 110.

67 *Ellerman Wilson Line v Webster* [1952] 1 Lloyd's Reports 179 (CA) 180 (Lord Goddard).

68 *Arrow Shipping Co v Tyne Improvement Commissioners, The Crystal* [1984] AC 508 (HL) 519 (Lord Herschell), 521 (Lord Watson), 527 (Lord Ashbourne), 534 (Lord Morris).

69 *Bradley v Newsom* [1919] AC 16 (HL) 24 (Lord Finlay).

70 Bridge, above n 51, at 23.

71 *Moorhouse v Angus and Robertson* [1981] 1 NSWLR 700 (NSW CA) 711.

72 554 A 2d 483 (Md App 1976).

73 554 A 2d 483 (Md App 1976) 499.

74 428 F 2d 683 (7th Cir 1970) 688.

75 McHale, above n 60, at 131.

(c) Remedies

In the event that the source's property rights have not been transferred to a third party, the next question is what remedies exist to protect interference with those property rights? In *Re Organ Retention Group Litigation*, Gage J considered the potential application of the tort of conversion to the removal of bodily materials from cadavers.[76] Property rights are protected in English law by common law and statute. In short, the common law protects property rights through the torts of trespass and conversion. Statutory protection exists under the Torts (Interference with Goods) Act 1977 UK (TIGA).

There are currently two torts in English law that protect property rights to chattels: trespass and conversion. Historically, there were three causes of action, but the tort of detinue was abolished by the TIGA.[77] The TIGA attempted to create a united regime for all interferences with goods through section 1, which provides that the collective phrase 'wrongful interference with goods' covers trespass, conversion, negligence and any other tort that results in damage to goods. The TIGA has not created a new tort of 'wrongful interference' in English law. The Act simply groups the common law torts together. Hence, the question of whether there has been a 'wrongful interference' in any given case remains a matter for the common law.[78]

The common law tort of conversion is the most useful vehicle for protecting property rights to chattels. Conversion is not, however, easily defined in the law.[79] In *OBG Limited v Allan; Douglas v Hello! Limited; Mainstream Properties Limited v Young*, Lord Hoffmann observed that '[a]nyone who converts a chattel . . . does an act inconsistent with the rights of the owner'.[80] The application of the tort of conversion raises two principal questions. The first concerns the level of interference with goods that is required for a claim in conversion. The second question lies in determining who is entitled to bring an action based on the tort of conversion. For the tort of conversion to be actionable, it is necessary that the claimant be denied the use and possession of the property in question.[81] Section 14 of the TIGA defines 'goods' as including 'all chattels personal other than things in action and money'. Tangible things that are not classified by the law as 'property' cannot be converted.[82] For example, conversion does not lie in respect of wild animals since

[76] *Re Organ Retention*, above n 32, is considered in detail in ch 2. See nn 69–96 and nn 181–91 of ch 2 and accompanying text.

[77] TIGA, s 2(1).

[78] A Tettenborn, 'Wrongful Interference with Goods' in AM Dugdale (ed), *Clerk and Lindsell on Torts* (London, Sweet & Maxwell, 19th edn, 2006) 1004–06.

[79] *Kuwait Airways Corporation v Iraqi Airways Co (No 4 and 5)* [2002] UKHL 19; [2002] 2 AC 883 (HL) [39] (Lord Nicholls).

[80] [2007] UKHL 21, [95]. Tettenborn describes conversion as 'an act of deliberate dealing with a chattel inconsistent with another's right whereby the other is deprived of the use and possession of it': Tettenborn, above n 78, at 1007.

[81] *Marq v Christie Manson & Woods Ltd (trading as Christie's)* [2004] QB 286, [13] (Tuckey LJ); *Kuwait Airways*, above n 79, at [38] (Lord Nicholls), [119] (Lord Steyn), [129] (Lord Hoffmann).

[82] J Davies, 'Tort' in P Birks (ed), *English Private Law* (Oxford, OUP, 2000) 503.

they are *res nullius*: no-one has a right to possession of a wild animal.[83] As Part I of this book demonstrated, the common law also considers that cadavers are not normally regarded as 'things' for the purposes of property law and so are not normally capable of being converted. This book contends that separated biological materials can be classified as 'things' which are, consequently, capable of conversion. If an object is classified as a 'thing', the question then becomes whether the claimant has been denied the use and possession of it. Common law authority has established that conversion will lie where a defendant: (1) takes possession of goods with no intention of returning them;[84] (2) refuses to hand over goods;[85] (3) delivers the goods into the hands of a third party;[86] (4) uses the goods in a manner inconsistent with the claimant's title;[87] or (5) destroys the goods.[88]

These forms of conversion are all potentially applicable in the context of separated biological materials. The fourth form is the most significant where separated biological materials are used for research without consent. For this category of conversion, the important legal issue is what degree of interference is required before the law will say that there has been a conversion. Common law authority treats the alteration of goods as conversion. Examples of conversion include a miller grinding corn into flour;[89] water being poured into wine;[90] and seals being cut from a deed.[91]

The second question in the context of conversion is: who is entitled to sue? To bring a claim for conversion, the claimant must have had possession or a right to immediate possession of the goods.[92] The Law Reform Committee (which was charged with preparing the materials for the TIGA) recognised the breadth of rights that can found an action for conversion, including ownership and possession. That committee noted that 'not only actual possession (or a right to immediate possession) at the material time, but also other interests in a chattel, whether present or future, possessory or proprietary . . . should constitute sufficient title to sue'.[93] In *Costello v Chief Constable of Derbyshire Constabulary*,[94] the Court of Appeal emphasised the possession-based nature of the action and held that the common law protects possession even where it has been obtained by theft or other unlawful means, as well as in the usual case of chattels lawfully obtained.[95]

Individuals are entitled to delivery up of goods and damages under section 3(2) of the TIGA. In the present context, a source may be entitled to damages if a

[83] Tettenborn, above n 78, at 1023.
[84] *Fouldes v Willoughby* (1841) 8 M & W 540, 549; 151 ER 1153, 1157.
[85] *Marq*, above n 81, at [33] (Tuckey LJ).
[86] *Consolidated Company v Curtis & Son* [1892] 1 QB 495, 498.
[87] *Mulgrave v Ogden* (1591) Cro Eliz 219, 219; 78 ER 475, 475.
[88] *Richardson v Atkinson* (1723) 1 Stra 576, 576; 93 ER 710, 711.
[89] *Hollins v Fowler* (1875) LR 7 HL 757 (Blackburn J).
[90] *Penfolds Wines Pty Ltd v Elliott* (1946) 74 CLR 204 (HCA) 229 (Dixon J).
[91] *Penfolds Wines*, ibid at 229 (Dixon J).
[92] Ibid; *The Winkfield* [1902] P 42 (CA) 55.
[93] Law Reform Committee, *Conversion and Detinue* (Cmnd 4774, 18th Report, 1971) 43.
[94] [2001] EWCA Civ 381; [2001] 1 WLR 1437 (CA).
[95] Ibid, at [31] (Lightman J).

recipient utilises biological materials: (1) without the consent of the source; or (2) goes beyond the terms of the bailment. In these circumstances, the tort of conversion protects unlawful interference with a source's property rights.

If a claim for conversion can be established, the claimant may then seek an order under the TIGA for delivery up of goods (at the court's discretion) and consequential damages; or an alternative order for delivery of the goods or damages, plus consequential damages; or damages only.[96] Damages may be awarded on the basis of either the claimant's loss or the defendant's gain. In *United Australia Ltd v Barclays Bank Ltd*,[97] the House of Lords established that the quantification of damages can proceed on either of these two bases. Compensatory damages for conversion aim to compensate for loss suffered and are, therefore, quantified by reference to the value of the chattel as at the date of the conversion.[98] The value of a chattel will generally be determined by reference to a market value (assuming a market exists).[99] This may be a difficult question in the context of biological materials because at the date of the conversion the biological materials may have limited commercial value. For example, it may be that some biological material only has value when used in combination with a large collection of biological materials contained in a biobank, or it may be that the value of the biological material lies in its use to create additional biological materials which may have substantial commercial value. The ownership of additional biological materials is considered below.[100]

If biological materials are utilised to make a profit, for example, in the development of additional biological materials, then the claimant would have a claim for the market value of the materials. If there is no market, then the damages can be measured by reference to a defendant's gain. Such awards are commonly referred to as restitutionary (and sometimes disgorgement) damages.[101] In *Attorney-General v Blake*, Lord Nicholls observed that awards of damages based on the defendant's gain had been made for claims involving an interference with property rights.[102] Further, an award of damages under the TIGA can include damages in respect of a defendant's gain.[103]

In addition to damages, a court may order the specific delivery of a chattel. Specific delivery is available under section 3 of the TIGA. At common law, an order for specific delivery is dependent on the nature of the object being delivered. The

[96] TIGA, s 3(2).

[97] [1941] AC 1 (HL).

[98] *BBMB Finance (Hong Kong) Ltd v Eda Holdings Ltd* [1990] 1 WLR 409 (PC) 412 (Lord Templeman); *General & Finance Facilities v Cook Cars (Romford) Ltd* [1963] 1 WLR 644 (CA) 649 (Diplock J).

[99] *Armory v Delamirie* (1722) 1 Stra 505.

[100] See nn 118–40 below and accompanying text.

[101] J Edelman, *Gain–Based Damages* (Oxford, Hart Publishing, 2002) 122.

[102] *Attorney-General v Blake (Jonathan Cape Ltd Third Party)* [2001] 1 AC 268, 278–81 (Lord Nicholls).

[103] *Hillesden Securities Ltd v Ryjak Ltd* [1983] 1 WLR 959 (CA) 963 (Parker J). Referred to with approval in *Kuwait Airways Corporation v Iraqi Airways Co* [2001] 3 WLR 1117 (CA) [521] (Brooke LJ).

common law draws a distinction between ordinary commercial objects and objects which are irreplaceable or peculiar in nature.[104] In the former case, the court is likely to make an award of damages based on the thing's market value.[105] The TIGA has not altered the common law position that a court will order specific delivery where the goods are unique or of some special value.[106] Given the personal nature of biological materials, it is likely a court would order specific delivery.

(d) Summary

Personal property principles (including those drawn from the law of bailment and abandonment) can be employed in the context of separated biological materials. Further, the TIGA and the law of conversion provide remedies for interference with any property rights. These principles offer a clear guiding framework.

2 Additional Biological Materials

Important questions concerning additional biological materials have not yet been addressed. In the factual scenario, B utilised X to create four other forms of additional biological materials: amplified DNA (X_1), cell-lines (X_2), modified biological materials (X_3), and a cell-line created from X_3 (X_4). Commentators and courts have assumed that additional biological materials are things which are brought within the law of property because of their physical existence. In *US v Arora*,[107] for example, the US District Court found that a cell-line was property capable of being converted. The court, however, was not required to examine the legal principle that explained the creation of property rights over the cell-line. The structural foundation of property rights is further developed here by discussing additional biological materials and examining applicable property principles to explain the creation and allocation of property rights.

(a) Medical and Scientific Research

Additional biological materials are increasingly important in the medical and scientific community as essential tools in the development of medical therapies and diagnostics.[108] The range of materials used include: 'cell lines, monoclonal antibodies, reagents, animal models, growth factors, combinatorial chemistry libraries, drugs and drug targets, clones and cloning tools (such as polymerase

[104] *Howard Perry Ltd v British Railways Board* [1980] 1 WLR 1375 (CA) 1382.

[105] A Burrows, *Remedies for Torts and Breach of Contract* (Oxford, OUP, 3rd edn, 2004) 578.

[106] *IBL Ltd v Coussens* [1991] 2 All ER 133 (CA) 139 (Neill LJ). For common law authority prior to the enactment of the TIGA, see: *Whiteley v Hill* [1918] 2 KB 808, 819.

[107] 806 F Supp 1091 (Md DC 1994).

[108] J Kahn, 'Law and Authority in Patenting Human Genetic Material' (2003) 14 *Stanford Law and Policy Review* 417, 434; HG Drexler, Y Matsuo and RAF MacLeod, 'Continuous Hematopoietic Cell Lines as Model Systems for Leukaemia-Lymphoma Research' (2000) 24 *Leukaemia Research* 881, 885.

chain reaction (PCR))'.[109] Progress in biomedical research is now more cumulative and depends more heavily on research tools already in existence.[110] Research tools are frequently transferred between researchers who use these tools to develop further materials. Often, researchers and institutions control the use of these materials through licence agreements and materials transfer agreements (MTAs).[111] Disputes will increasingly arise between researchers, institutions and biotechnology companies over the legal status of these materials and the property rights attaching to them.

(b) Intellectual Property

Questions as to the intellectual property rights attaching to such materials are also likely to arise. In some instances, intellectual property rights will attach to these materials as 'inventions'. There is an ever-expanding collection of secondary literature that considers the scope of intellectual property rights over biological materials and genetic information.[112] Questions of intellectual property rights have also attracted the attention of law reform bodies.[113] These law reform bodies, however, have not considered what underlying property rights are applicable to these additional biological materials. This is primarily because the commercial interest in these materials often lies in licensing the use of inventions, which are protected through intellectual property rights. Ownership of the physical materials is, thus, not the focus of protection.

The opportunity to consider an analogous question arose in a recent Canadian case—*Monsanto Canada Inc v Schmeiser*[114]—where the Canadian Supreme Court considered the appropriate intellectual property protection for a molecularly engineered gene. In that case, Monsanto accused Schmeiser (a farmer) of making, using and selling its patented invention (glyphosate-resistant canola seed) without a licence. Monsanto's private investigators discovered traces of their canola seed in Schmeiser's crop. Schmeiser claimed that the seed had blown onto his land and he had inadvertently planted the seed. Thus, the case hinged on a potential clash

[109] National Institutes of Health, *Report of the National Institutes of Health (NIH) Working Group on Research Tools* (1998) (available at http://www.nih.gov/news/researchtools, accessed on 1 June 2007).

[110] JP Walsh, WM Cohen and A Arora, 'Licensing Research Tools and Biomedical Innovation' in S Merrill, R Levin and M Meyers (eds), *Innovation in a Knowledge Based Economy* (Washington, National Academy Press, 2003) 289.

[111] AL Monotti, 'Access to Tangible Research Materials in Biomedical Research: Conditions of Access and their Effect on Research' (2006) 14 *Journal of Law and Medicine* 86, 87.

[112] See, for example: D Nicol, 'On the Legality of Gene Patents' (2005) 29 *Melbourne University Law Review* 809; A Warren-Jones, 'Patenting DNA: A Lot of Controversy Over a Little Intangibility' (2004) 12 *Medical Law Review* 97; Kahn, above n 108; RS Eisenberg, 'Patents on DNA Sequences: Molecules and Information' in N Elkin-Koren and NW Netanel (eds), *The Commodification of Information* (The Hague, Kluwer Law International, 2002) 415.

[113] Australian Law Reform Commission, *Gene Patenting and Human Health* (ALRC 99, Australian Government Publishing Service, Canberra, 2004); Nuffield Council on Bioethics, *The Ethics of Patenting DNA* (Nuffield Council on Bioethics, London, 2002).

[114] [2004] SCC 34; [2004] 1 SCR 902 (SCC).

between the company's intellectual property rights and the underlying personal property rights of a farmer over canola seeds.

On appeal, the Canadian Supreme Court considered the question before the court to be one of patent protection rather than ownership of the seeds. For the Supreme Court, ownership of the seeds was not a defence to a breach of the Patents Act.[115] In this way, the Supreme Court collapsed the questions of rights concerning the intellectual property and of personal property rights over the physical property.[116] The conflation of these issues is similar to the approach adopted by the majority of the Californian Supreme Court in *Moore*. By relying on statutory intellectual property rights for their arguments, both courts ignored the potential property rights existing over the biological materials (in the case of *Moore*) and the plant seed (in the case of *Schmeiser*). Questions over the legal status of the physical materials are, however, separate questions from those relating to patent protection.

Despite the approach of the Canadian Supreme Court in *Schmeiser*, the legal status of, and property rights attaching to, these additional biological materials are potentially significant issues. For example, there may be no intellectual property rights over a particular form of biological material that is being used.[117] Further, even if intellectual property rights exist over additional biological materials, such rights do not determine to whom property rights over the physical materials are allocated. It is still necessary to consider what property principles are applicable in relation to the physical status of additional biological materials.

(c) Specification and Accession

Specification and accession are the applicable property law doctrines to determine the creation and allocation of property rights to additional biological materials. Chapter five noted the distinction between the doctrines of specification and accession, on the one hand, and the rules governing mixing on the other.[118] Importantly, the rules governing mixing are only applicable in the context of divisible mixtures such as oil or grain. For mixtures with an indivisible joinder, then the applicable principles are accession and specification.[119] In the factual scenario presented (concerning X_1, X_2, X_3 and X_4), the facts involve an indivisible joinder. Accordingly, the rules governing mixing are not relevant for the discussion here.

In considering the application of the accession and specification doctrines to additional biological materials, it is first necessary to determine whether these doctrines can apply in the case of natural matter. There is no judicial authority in England or Australia on this point, and in the US, *US v Arora*[120] stands as the sole

[115] *Schmeiser*, ibid at [96].

[116] J DeBeer, 'Reconciling Property Rights in Plants' (2005) 8 *Journal of World Intellectual Property* 5, 5.

[117] Walsh, Cohen and Arora, above n 110, at 319.

[118] See nn 46–58 of ch 5 and accompanying text.

[119] L Smith, *The Law of Tracing* (Oxford, Clarendon Press, 1997) 70.

[120] 806 F Supp 1091 (Md DC 1994).

decision addressing the issue. The Roman jurists did not consider the question of property rights to biological materials, because at that time such materials had no commercial value. Instead, the issues of the day concerned rights over things such as the young of animals.[121] One commentator has observed that 'there may be an argument for specification in the case of the development of cells from matter taken from a living organism'.[122]

A case in Scotland—*Kinloch Damph Ltd v Nordvik Salmon Farms Ltd*[123]—addressed the legal status of a natural substance and serves as a useful example of applying the specification principle (*specificatio*) to living natural substances. In *Kinloch*, the pursuers sold 1.25 million smolt (young salmon in embryonic state) to the defenders. Under the contracts, title did not pass to the defenders until the price was paid. The retention of title clause provided that '[u]ntil the price for the goods is paid for in full, the goods shall, notwithstanding delivery, remain the property of the seller'. The defenders placed the smolt into cages in seawater and fed and raised them. The fish developed into salmon thirty times their original size. The defenders went into receivership and there was an outstanding balance of approximately £700,000 for the smolt. The pursuers brought proceedings for the return of the salmon, claiming property rights in respect of the salmon by virtue of their reservation of title clause. The question for the Scottish court lay in deciding whether the salmon as they now existed were (by virtue of the retention of title clause) the property of the pursuers, or instead, the property of the defenders.

The defenders advanced two arguments. First, they argued that the retention of title clauses constituted an ineffective attempt to create a security over goods in the possession of the defenders. Secondly, the defenders argued that the smolt had ceased to exist after the addition of feeding and husbandry, such that the salmon in the defenders' possession were new things. Thus, they argued that property rights should be allocated to them under the specification doctrine. In response, the pursuers submitted that the circumstances in question had not created a new thing, and that the doctrine of *specificatio* could not be displaced by contract.

Lord Macfadyen held that the retention of title clauses were valid and that the salmon remained the property of the pursuers.[124] He concluded that a *nova species* was not created, noting that smolt and salmon are the same biological species.[125] Lord Macfadyen observed that the development of the smolt into mature salmon was irreversible. He also argued that the specification doctrine should be limited to inanimate objects or substances created by human effort that are used up and cease to exist in the process of creation. Lord Macfadyen's principal justification for limiting the application of specification concerned the lack of authority deter-

[121] P Stein and WW Buckland, *A Textbook of Roman Law* (Cambridge, CUP, 3rd edn, 1968) 222–23.

[122] DL Carey Miller, *Corporeal Moveables in Scots Law* (Edinburgh, W Green, 2nd edn, 2005) 91. See, also: D Johnston, 'The Renewal of the Old' [1997] 56 *Cambridge Law Journal* 80, 91–94.

[123] Unreported 30 June 1999 (Outer House Court of Session) (available at http://www.scotcourts.gov.uk/opinions/ca291499.html, accessed on 1 June 2007).

[124] *Kinloch*, ibid at 7.

[125] JW Jones, *The Salmon* (London, Collins, 1959) 153–56.

mining the application of the doctrine to the natural growth of living creatures. Significantly, Lord Macfadyen noted:

> The examples in the writing on this subject contain no references to specification of growing animals. I consider that there is force in the submission that, having regard to the much greater importance of animals in daily life in former times (whether the times of the Roman writers or those of the Scottish institutional writers), the absence of such reference is a strong indication that the doctrine had no such application.[126]

On this basis, Lord Macfadyen concluded that mature salmon were, 'as objects of ownership', the same things as the smolt. The pursuers, therefore, retained title to the salmon. Lord Macfadyen was correct in concluding that the specification doctrine should not have been applied. Chapter five surmised that English law considers a new thing to be created where the original material ceases to exist and can no longer be restored to its original physical state.[127] In the case of the *natural* growth of living creatures, the original material still exists, albeit in a more mature form. Therefore, in the cases of natural growth where the creature or living matter develops in a way as intended by nature, the specification doctrine has no application.

Lord Macfadyen's broader claim that the specification doctrine cannot apply to natural substances does not necessarily follow. The more difficult question pertaining to natural substances is what happens if the natural substance does not mature by way of natural growth or decay. Intervention through human endeavour may, for example, affect the natural growth or decay process. The fact that Roman texts did not refer to examples involving living animals is not, in itself, a substantial reason to refuse extension of the specification doctrine to living material. Moreover, there are plausible explanations as to why the Roman texts do not provide examples. Most obviously, it is conceivable that disputes simply did not arise concerning the transformation of living substances into new things. The state of industry at that time did not require the Roman jurists to consider such issues, for they were concerned with industrial and agricultural processes.

Secondly, if a dispute arose concerning the ownership of living or growing items, the rules on the acquisition of fruits would have provided a satisfactory answer to issues at that time.[128] The 'general fruits principle' provides that all natural products of the animal and plant kingdoms—including the milk of a cow or apples on a tree—are accessories of the animals or plants from which they derive.[129] The owner of the animal or tree also acquires ownership of the accessory. As a straightforward example, if an individual owns an apple tree then, under Roman principles, that individual also owns all the apples that it produces. On the

126 *Kinloch*, above n 123, at 7.

127 See nn 63–88 of ch 5 and accompanying text.

128 E Metzger, 'Postscript on *nova species* and *Kinloch Damph Ltd v Nordvik Salmon Farms Ltd*' (2004) 2 *Roman Legal Tradition* 115, 118.

129 GJ Bell, *Principles of the Law of Scotland* (Edinburgh, Butterworths, 10th edn, 1899) 503; B Nicholas, *An Introduction to Roman Law* (Oxford, Clarendon Press, 1962) 138; KGC Reid, *The Law of Property in Scotland* (Edinburgh, Butterworths, 1996) 595.

other hand, if an ancient Roman had developed a way to make another's apple trees grow pears, and then demanded ownership of the pears, an application of the 'accession-by-fruits principle' may not produce a just result. An application of the principle would determine that the owner of the apple trees would be entitled to the pears as well, despite the work performed by the person who had developed the technique to grow pears.

The more difficult question, then, is whether specification can apply where natural substances do not mature by a matter of natural growth or decay, but instead, through human intervention. In *Kinloch*, Lord Macfadyen did not consider the relevant principle where a natural growth process becomes altered following human intervention. The specification doctrine is capable of application, at least in principle. The introduction of human endeavour to biological materials is analogous to the human endeavour recognised by the Roman jurists in relation to agricultural processes; it is simply a more advanced form of industrial application. Where the natural pattern of growth or decay is altered, then the original material ceases to exist and a new thing is created. In the factual scenario presented in this chapter, B intervenes in the natural decay of X through the creation of amplified DNA, cell-lines and modified biological materials. The question of whether a new thing is created in each of these cases is discussed below. However, if a new thing can be said to have been created, then under the specification doctrine, B should be allocated the property rights.

At present, no English authority has employed the accession or specification doctrines with respect to a natural growth or decay process. There is, however, a principle at common law dictating that offspring (with the exception of swans) belong to the owner of the female where the owner has the female in possession.[130] The principle represents an application of the maxim *partus sequitur ventrem* (the offspring follow the dam), and such cases are classified as examples of accession.[131] This principle has been applied to 'stray bull cases', which concern the ownership of progeny where a bull belonging to the defendant trespasses onto the plaintiff's land and impregnates the plaintiff's cows. There is no direct authority in England addressing such a scenario, although the English Court of Appeal has acknowledged the application of the accession principle in the 'stray bull' context.[132] The principle has also been recognised in academic commentary.[133] Further, there are cases applying the principle in other jurisdictions.[134] These cases, in effect, adopt the accession doctrine with respect to the offspring of live animals. They reason that, as a matter of biology, the sperm and ovum are combined to form a progenitor cell at the initial stage. The bull provides material that ultimately constitutes a small but innate part of the offspring, and, therefore, accedes to the principal, the

[130] Blackstone, above n 64, at vol II, 391; *Case of Swans* (1592) 7 Co Rep 15b, 17ba; 77 ER 435, 438.

[131] Tyler and Palmer, above n 65, at 430.

[132] *Tucker v Farm and General Investment Trust Ltd* [1966] 2 QB 421, 426, 430.

[133] Bridge, above n 51, at 107; AG Guest, 'Accession and Confusion in the Law of Hire Purchase' (1964) 27 *Modern Law Review* 505, 506–07.

[134] *Popowich v Letweniuk* [1972] 1 WWR 641 (Sask DC); *Neeb v Hoffman* [1989] OJ No 302 QL (Ont DC).

cow. This English law principle highlights that the common law can employ and apply Roman law principles to natural growth or decay processes.

(d) Application

The principles of specification and accession assist in determining the creation and allocation of property rights to amplified DNA, cell-lines and modified biological materials. These are important questions because these materials will increasingly become the subject of disputes, not just between sources and researchers, but between researchers, universities and biotechnology companies as well. The legal question in each case is whether a new thing is created in law. English law recognises that a new thing is created where the original material ceases to exist and can no longer be restored to its original physical state.[135] In the alternative, the principle of accession applies where there is a principal and subordinate item and the subordinate item is subsumed into the principal.[136]

The first form of additional biological material to be addressed is amplified DNA (X_1). In order to determine whether specification or accession is the better governing principle to be applied, it is necessary to understand the physical composition of X_1. When DNA is amplified, a copy of the original DNA is made in a test tube (*in vitro*). The chemicals that make up the new biological structures of X_1 are added and do not come from the original DNA. X_1 is, therefore, a chemical copy of the original sample. A tiny amount of the original cellular DNA remains present in X_1 because the original physical material was used as the template for making the copies. In a typical amplification, about 10 nanograms of human DNA might be amplified to about 100–500 micrograms of DNA, which is about a 10,000–50,000 fold amplification.[137] Approximately 1/10,000th–1/50,000th of the amplified DNA will consist of the original DNA from the cell. The rest of X_1 comprises new materials supplied by B.

Scientific analysis indicates that the original DNA acts as a subordinate item to the principal entity, which consists of the physical material in the amplified DNA. It is unlikely, therefore, that the law would view X_1 as a new thing. It would be more accurate to view the original DNA as acting as a subordinate item that is subsumed into the new materials. The better analogy may lie, instead, with the doctrine of accession, and, at this point, it is useful to recall the stray bull cases. Those cases provide a good analogy, illustrating how the common law can apply the principle of accession in the context of living matter. Under the principle of accession, then, the property rights to X_1 are allocated to B as the owner of the new materials that are added.

A similar analysis applies to cell-lines (X_2). A cell-line is derived from a single cell. When that cell divides there are two cells, each of which, prior to the division,

135 See nn 63–88 of ch 5 and accompanying text.

136 See nn 50–58 of ch 5 and accompanying text.

137 RS Lasken and M Egholm, 'Whole Genome Amplification: Abundant Samples of DNA from Precious Samples or Clinical Specimens' (2003) 21 *Trends in Biotechnology* 531, 532.

was physically part of the first cell. Further division creates four cells. Two of these cells were physically part of one of the first of the two cells; the other two were physically part of one of the second of the two cells. The later cells in the cell-line could be described as direct physical descendents of the original cell. As with amplified DNA, a mix of organic materials (including nutrients) must be added to ensure that X_2 can continue to self-replicate.[138] A new thing is arguably created and the doctrine of specification should apply. However, this process is similar to the amplification of DNA, and accession offers a more appropriate analysis. As with X_1, B would still be entitled to the property rights to X_2 because the mix of organic materials would constitute the principal item. In the event that the law considers a new thing to have been created, then B is still entitled to the property rights to X_2 as the maker of the new thing.

The third example is modified biological materials (X_3). Chapter one detailed how foreign DNA is inserted into original biological materials and cell-lines.[139] Foreign DNA is frequently used in the context of immunology. In the case of X_3, the legal question once again becomes whether (following the insertion of foreign DNA) a new thing is created, or whether an addition to X has simply been made. This is a difficult question: it could be, for example, that the modification to X is so insubstantial that it is merely an accession. If so, the property rights to X_3 are not allocated to B, but rather to A as the original owner of X. Alternatively, it may be that a new thing is created. For example, if B inserts foreign DNA into X, it may substantially alter the genetic composition of the modified biological materials. In the example of modified biological materials discussed in chapter one, the function of the cell can change following the insertion of foreign DNA. Arguably, the nature of the original biological materials has changed so substantially that the law can say that a new thing has been created. Yet, although the nature of the cell has changed, it remains a cell that functions as a cell. As a result, it may be difficult to argue that, in law, a new thing exists.

Instead, the more persuasive characterisation of X_3 is that two different materials (the original biological materials and the foreign DNA) have been joined together. Accordingly, the law of accession should again be the governing property principle. In these circumstances, the secondary question is whether the original biological materials, or the foreign DNA inserted into them, should be considered the principal entity. In the context of accession, the foreign DNA has become the dominant entity and so the property rights to the modified biological materials should be allocated to B. Such a characterisation is dependent on the characterisation of the foreign DNA as the principal entity. This may be problematic because, although the foreign DNA has the effect of altering the nature of the original biological material, it only forms a small physical component of the modified biological material. To date, the common law has focused on the physical composition of materials. Here, it is likely that the original biological materials will remain the

[138] GM Cooper, *The Cell: A Molecular Approach* (Massachusetts, Sinauer, 3rd edn, 2004) 31.
[139] See n 67 of ch 1 and accompanying text.

principal entity and the initial property rights in X_3 will be allocated to A. Chapter one noted that modified biological materials are often used in scientific research to create cell-lines.[140] The modified biological material (X_3) is used as the original material to create the cell-line. In the event that X_3 is used to create a cell-line (X_4), then the analysis conducted above in relation to cell-lines applies, and the property rights to X_4 should be allocated to B.

(e) Summary

The doctrines of accession and specification are useful legal principles in determining the creation and allocation of property rights over additional biological materials. Although difficult factual scenarios remain, employing these doctrines offers a more principled approach than that adopted by the Californian Supreme Court majority in *Moore*.

C Conclusion

This chapter commenced by examining two potential explanations for the creation of property rights. First, the physical act of detachment was considered. The 'detachment principle' provides a clear and forceful explanation for the creation of property rights. English law may remain reluctant to recognise detachment as sufficient to create property rights. Accordingly, the chapter explored whether property rights could be created where an intention is also expressed for the separated biological materials to be used as property. The 'detachment plus intention principle' provides a further explanation for the creation of property rights in the event that the source expresses the requisite intention prior to, or on, detachment of the biological materials. The chapter outlined the conceptual difficulties that arise where a person other than the source expresses the requisite intention, and then suggested principles to address these difficulties. Importantly, the chapter concluded that the property rights created under the 'detachment principle' or 'detachment plus intention principle' should be allocated to the living person from whom the biological materials were removed.

This chapter then placed the doctrinal and theoretical analysis in a practical context by applying a property-based approach to a more complicated factual scenario arising in the context of scientific research. First, applicable legal principles concerning the transfer and abandonment of property rights by the source were examined. Secondly, the analysis focused on the creation and allocation of property rights to additional biological materials by applying the specification and accession doctrines. An understanding of these doctrines is necessary to deal with future disputes over such materials.

[140] See nn 67–69 of ch 1 and accompanying text.

There is a tide running against the recognition of property rights to separated biological materials in English law. This is a tide of policy concerned with the right of an individual to sell separated biological materials. Fears over the commodification of the human body do present substantial obstacles to the recognition of property rights in English law. Part I noted that the law has also invoked non-proprietary interests to protect an individual's autonomy and human dignity. Consequently, chapter seven examines the potential protection of an individual's *non-proprietary* interests in separated biological materials.

7

Non-proprietary Interests

One of the themes identified in Part I was that the law has sought to protect individual autonomy and a respect for human dignity. Chapter two demonstrated that US jurisprudence, in particular, has developed such protection in the context of dead bodies through the law of tort.[1] It is now necessary to consider whether English law can recognise and protect an individual's (or those representing the deceased) *non-proprietary* interests in separated biological materials. Three issues are considered here. First, this chapter identifies what non-proprietary interests an individual (and those connected to a deceased person) can assert in relation to separated biological materials. Secondly, it examines whether the English law of wrongs can develop so as to protect these interests. Finally, this chapter considers whether English law is open to the development of liability for the non-consensual use of separated biological materials.

A Protected Interests

The first step in considering non-proprietary interests, as identified by Lord Steyn in *Chester v Afshar*, is to 'identify precisely the protected legal interests at stake'.[2] A living person has two interests that English law could protect in respect of separated biological materials: (1) autonomy; and (2) freedom from emotional distress. In the context of cadavers, the relatives of the deceased also have two interests: (1) respect for private and family life as part of the broader concept of autonomy; and (2) freedom from emotional distress.

First, a living person may claim a right to retain control of separated biological materials to ensure respect for their individual autonomy. Two developments in English law suggest that living persons retain an interest to control separated biological materials based on a respect for individual autonomy.

The first development is located in English statutes that require an individual's consent for the use of separated biological materials. Chapter three outlined these

[1] See nn 222–34 of ch 2 and accompanying text.

[2] [2004] UKHL 41; [2005] 1 AC 134 (HL) [18]. See, also: *Harriton (by her tutor Harriton) v Stevens* [2006] HCA 15; (2006) 226 CLR 52 (HCA) [225] (Crennan J); *Cattanach v Melchoir* [2003] HCA 38; (2003) 215 CLR 1 (HCA) [23] (Gleeson CJ).

statutory regimes.[3] More recently, the Human Tissue Act 2004 UK (HTA 2004) has established that a living person's consent (or consent from an appropriate representative) is required for the use of relevant material. Under the HTA 2004, a criminal offence is committed where appropriate consent is not obtained for the use of relevant material.[4] The HTA 2004 also establishes a separate offence relating to non-consensual testing of DNA derived from bodily materials.[5] The creation of these offences demonstrates that English law-makers have already recognised (at least to some extent) that individuals maintain an interest in, and a right to control, separated biological materials. Indeed, the right that individuals have to control such materials was emphasised by the government throughout the parliamentary process.[6]

The second development in English law is found in the dicta of Lord Woolf CJ and Baroness Hale in *R (S) v Chief Constable of South Yorkshire Police; R (Marper) v Chief Constable of the South Yorkshire Police (R (S))*.[7] Lord Woolf CJ (in the Court of Appeal) considered that separated biological materials are 'personal' to the individual.[8] In the House of Lords, Baroness Hale observed that individuals have privacy interests over separated biological materials similar to those over genetic information extracted from such materials.[9] While both comments were *obiter*, they acknowledge that a living person may have non-proprietary interests in separated biological materials that are worthy of legal recognition and protection.

In the context of dead bodies, English law also recognises the right of relatives of a deceased person to respect for private and family life. In *Re Organ Retention*, Gage J found that the non-consensual use of a deceased child's brain violated article 8 of the European Convention on Human Rights (ECHR), where the consent of the child's parents had not been obtained.[10] A right to respect for private and family life fits within the rubric of autonomy and will be examined in this chapter under that interest.[11]

The second interest that English law can protect with respect to separated biological materials is the right to freedom from emotional distress. There may be circumstances where individuals suffer emotional distress following the non-consensual use of separated biological materials. This could happen, for example, in the event that they discover their biological materials have been used to create cell-lines, or for research in which they did not wish to participate. In the context

[3] See nn 96–115 of ch 3 and accompanying text.

[4] HTA 2004, s 5.

[5] HTA 2004, s 45.

[6] See, for example: HC Standing Committee G, cols 065, 066 and 068 (27 January 2004). In the Commons Committee Debates, Dr Ladyman stated that: '[t]he right way forward is a statutory amendment with a golden thread running through it stating that people have the right to be consulted about the use of material from their own bodies' (col 068).

[7] [2002] EWCA Civ 1275; [2002] 1 WLR 3223 (*R (S)* CA); and [2004] UKHL 39; [2004] 1 WLR 2196 (*R (S)* HL).

[8] *R (S)* CA, ibid, at [32].

[9] *R (S)* HL, above n 7, at [70].

[10] *Re Organ Retention Group Litigation* [2004] EWHC 644; [2005] QB 506, [288]–[89].

[11] *R (Burke) v General Medical Council* [2004] EWHC 1879; [2005] QB 424, [59] (Munby J).

of dead bodies, the relatives of the deceased person also have a right to freedom from emotional distress. The inquiries into post-mortem practices at the Bristol Royal Infirmary, the Royal Liverpool Children's Hospital (Alder Hey) and other NHS hospitals during the late 1980s and early 1990s demonstrate the impact that the non-consensual removal of biological materials can have on family members of the deceased person.[12]

B English Law

Part II established that English law does not currently protect an individual's non-proprietary interests in separated biological materials. Even though the possibility of protection was raised by Baroness Hale in *R (S)*, English courts have not yet examined the scope of an individual's non-proprietary interests. Further, chapter three concluded that separated biological materials fall outside the protection afforded by the Data Protection Act 1998 UK (DPA).[13] The analysis in this chapter takes the interests of autonomy, and freedom from emotional distress, and considers how English law currently protects them. It then examines the potential extension of existing wrongs in relation to separated biological materials.

1 Autonomy

Autonomy and respect for human dignity are individual interests that are capable and worthy of protection in English law.[14] Chapter one defined and analysed these interests as they have been developed and employed in English law to date. 'Human dignity' was defined as the intrinsic worth of the individual, and 'autonomy' as the capacity that competent individuals have to act independently.[15] In *McFarlane v Tayside Health Board*, Lord Millett noted that autonomy can be viewed as an aspect of human dignity and that the protection of autonomy can also be viewed as the protection of human dignity.[16]

[12] Bristol Royal Infirmary Inquiry, *Report of the Royal Liverpool Children's Inquiry* (2001) (available at http://www.bristol-inquiry.org.uk/index.htm, accessed on 1 June 2007); Redfern Inquiry into the Royal Liverpool Children's NHS Trust–Alder Hey (2001) (available at http://www.rlcinquiry.org.uk, accessed on 1 June 2007); Independent Review Group, *Retention of Organs at Post-Mortem in Scotland* (2001) (available at http://www.show.scot.nhs.uk, accessed on 1 June 2007).

[13] See nn 220–40 of ch 3 and accompanying text.

[14] The latest edition of *Clerk and Lindsell on Torts* lists autonomy as an individual interest that the English law of torts protects: AM Dugdale, 'Principles of Liability in Tort' in AM Dugdale (ed), *Clerk and Lindsell on Torts* (London, Sweet & Maxwell, 19th edn, 2006) 21. See, also: J Davies, 'Tort' in P Birks (ed), *English Private Law* (Oxford, OUP, 2000) 509. Note, however, that some consider autonomy is not an interest traditionally protected by the tort of negligence: J Stapelton, 'Occam's Razor Reveals an Orthodox Basis for *Chester v Afshar*' (2006) 122 *Law Quarterly Review* 426, 442.

[15] See nn 110–11 of ch 1 and accompanying text.

[16] [2000] 2 AC 59 (HL) 114.

(a) Underlying Value

It is first necessary to develop an understanding of the intersecting interests of autonomy and human dignity in English law. In other jurisdictions, human dignity is extremely important. Under German law, for example, human dignity is enshrined explicitly in the German Constitution.[17] It is not yet clear what role autonomy and human dignity will assume in English law. Birks has argued that a respect for human dignity operates as an umbrella concept for the wrongs that protect various aspects of personal dignity.[18] According to this view, human dignity operates as a general principle protecting a number of sub-interests, such as privacy and reputation. This view has not been accepted by English courts.

Historically, English law has been reluctant to grant remedies for infringements on autonomy and human dignity. These interests are increasingly protected under English law, however, following incorporation of the ECHR.[19] Article 8(1) of the ECHR provides, inter alia, that each person has 'the right to respect for his private and family life'. It is important to note, first, the approach of the ECtHR to article 8, because of the potential impact that the ECHR and ECtHR's jurisprudence may have on the development of English law.[20] In broad terms, the ECtHR has held that the right to respect for private and family life protects both physical and moral integrity.[21] According to the ECtHR, moral integrity is related to human dignity and the protection of moral integrity is intended to ensure the development 'of the personality of each individual in his relations with other human beings'.[22] The ECtHR considers autonomy and human dignity to be important, integral aspects of article 8.

Even though autonomy and human dignity are often considered elusive concepts, English law has not ignored these European developments. In *Evans v Amicus Healthcare Ltd*, Arden LJ observed:

> The fact is that each person has a right to be protected against interference with their private life. That is an aspect of the principle of self-determination or personal autonomy.[23]

Questions regarding the place of autonomy and human dignity in English law have arisen primarily in relation to arguments concerning the existence of a right

[17] *Grundgesetz* (GG) (German Federal Constitution), art 1. One author has predicted that the German jurisprudence interpreting art 1 will influence the development of European law: J Jones, 'Common Constitutional Traditions: Can the Meaning of Human Dignity Under German Law Guide the European Court of Justice' [2004] *Public Law* 167, 183.

[18] P Birks, 'Harassment and Hubris: The Right to an Equality of Respect' [1997] *Irish Times* 1, 44. This argument was first made in a North American context: EJ Bloustein, 'Privacy as an Aspect of Human Dignity: An Answer to Dean Prosser' (1964) 39 *New York University Law Review* 962, 1003.

[19] *Rees v Darlington Memorial Hospital NHS Trust* [2003] UKHL 52; [2004] 1 AC 309 (HL) [123] (Lord Millett); J Wright, *Tort Law and Human Rights* (Oxford, Hart Publishing, 2001) 14.

[20] Section 2(1)(a) of the Human Rights Act 1998 UK requires English courts to take relevant ECtHR jurisprudence into account when determining any question arising in connection with an ECHR right.

[21] *Costello–Roberts v United Kingdom* (1993) Series A, No 247-C; (1995) 19 EHRR 112.

[22] *Case of Von Hannover v Germany* (2005) 40 EHRR 1, [50]; *Botta v Italy* (1998) 26 EHRR 241, [32].

[23] [2004] EWCA Civ 727; [2005] Fam 1 (CA) [110].

to privacy. Privacy is a highly contentious and widely debated concept. It is defined broadly here as the 'personal space' that an individual is entitled to control.[24] The Canadian Supreme Court has observed that privacy interests include 'personal privacy, territorial privacy and informational privacy'.[25] The English duty of confidence is confined to informational privacy, which includes the right that individuals have to control personal information about them.[26] Since the publication by Warren and Brandeis of their seminal article, 'The Right to Privacy',[27] the question of the existence of a privacy tort within the common law has been debated throughout the common law world. American courts have embraced a tort of privacy,[28] as has the New Zealand Court of Appeal.[29] In contrast, the High Court of Australia,[30] and the Supreme Court of Canada,[31] left open the question of whether such a tort exists.

Traditionally, English courts have approached the field of privacy with caution and reticence. Following the incorporation of the ECHR into English law, in *Douglas v Hello! Ltd*,[32] Sedley LJ observed that the law can recognise 'privacy itself as a legal principle drawn from the fundamental value of personal autonomy'.[33] In *Wainwright v Home Office*,[34] the House of Lords rejected this contention, and concluded that a general privacy tort had not emerged in English law following the incorporation of the ECHR.[35] *Wainwright* concerned an unauthorised strip-search of Mrs Wainwright and her son while visiting a relative in prison. The prison officers failed to follow correct procedures, thereby exceeding their

[24] *R v Broadcasting Standards Commission; ex parte BBC* [2001] QB 885, [48] (Lord Mustill).

[25] *R v Tessling* [2004] SCC 67; (2004) 244 DLR (4th) 541 (SCC) [20].

[26] *Tessling*, ibid at [23].

[27] SD Warren and LD Brandeis, 'The Right to Privacy' (1890) 4 *Harvard Law Review* 193.

[28] One author claims that most, if not all, US jurisdictions have acknowledged a common law right to privacy in one form or another: G Deny, 'The Newsworthiness Defense to the Public Disclosure Tort' (1996) 85 *Kentucky Law Journal* 147, 147. The US common law tort of privacy is generally divided into four parts: (i) intrusion on the plaintiff's solitude or seclusion; (ii) public disclosure of embarrassing facts about the plaintiff; (iii) publicity putting the plaintiff in a false light; and (iv) appropriation of the plaintiff's name or likeness for private advantage. These categories follow the classification provided by WL Prosser, 'Privacy' (1960) 48 *California Law Review* 383.

[29] *Hosking v Runting* [2004] NZCA 34 (NZ CA) [118] (Gault P and Blanchard J), [248] (Keith J), [263] (Anderson J).

[30] *Australian Broadcasting Corporation v Lenah Game Meats* [2001] HCA 63; (2001) 208 CLR 199 (HCA) [41]–[43] (Gleeson CJ), [132] (Gummow and Hayne JJ), [188]–[89] (Kirby J), [335] (Callinan J). Since *Lenah Game*, two Australian judges have held that the common law in Australia recognises a tort of privacy: *Jane Doe v Australian Broadcasting Corporation & Ors* [2007] VCC 281 (County Court Vic) [157] (Hampel J) and *Grosse v Purvis* [2003] QDC 151 (Qld DC) [442] (Skoien SJ). Note, however, that the Full Court of the Federal Court of Australia has observed that Australian law has not developed to the 'point where an action for breach of privacy is recognised': *Kalaba v Commonwealth of Australia* [2004] FCAFC 326, [8] (Tamberlin, North and Dowsett JJ).

[31] *Les Editions Vice-Versa Inc v Aubry and Canadian Broadcasting Corporation* (1998) 157 DLR (4th) 577 (SCC) [6] (Lamer CJ).

[32] [2001] QB 967.

[33] *Douglas*, ibid at [126].

[34] [2003] UKHL 53; [2004] 2 AC 406 (HL).

[35] *Wainwright*, ibid at [1] (Lord Bingham), [30] (Lord Hoffmann), [54] (Lord Hope), [55] (Lord Hutton), [62] (Lord Scott).

statutory authority. The claimants advanced two arguments. The first was based on the principle in *Wilkinson v Downton*,[36] which, the claimants argued, provided a cause of action for the intentional infliction of emotional distress. The *Wilkinson* principle is considered in more detail below. In the alternative, the claimants submitted that English law should recognise a general tort of invasion of privacy. The House of Lords rejected both submissions.

Delivering the leading speech, Lord Hoffmann opined that English law protects the privacy interests of an individual through the (much expanded) duty of confidence.[37] He suggested that, in *Douglas*, Sedley LJ had not advocated 'a high-level principle of invasion of privacy'.[38] Instead, in an important passage Lord Hoffmann stated:

> There seems to me a great difference between identifying privacy as a value which underlies the existence of a rule of law (and may point the direction in which the law should develop) and privacy as a principle of law in itself. The English common law is familiar with the notion of underlying values—principles only in the broadest sense—which direct its development.[39]

Lord Hoffmann also acknowledged the wider influence in English law of the privacy 'value':

> There are a number of common law and statutory remedies of which it may be said that one at least of the underlying values they protect is a right to privacy . . . Common law torts include trespass, nuisance, defamation and malicious falsehood; there is the equitable action for breach of confidence and statutory remedies under the Protection from Harassment Act 1997 and the Data Protection Act 1998.[40]

From these passages, it is clear that the concept of privacy is an underlying value in English law. *Wainwright* was taken to the European Court of Human Rights, and the Court found that the House of Lords' narrow view of privacy provided insufficient protection for the applicant's rights under article 8 of the ECHR and accordingly awarded damages.[41] Nevertheless, as the law in England currently stands, privacy is not considered a separate right capable of direct application. In *Wainwright*, Lord Hoffmann noted that 'English law has so far been unwilling, perhaps unable, to formulate any such high-level principle'.[42] Similarly, Lord Nicholls confirmed in *Campbell* the refusal of English law to develop an 'overarching, all-embracing cause of action for "invasion of privacy"'.[43] Instead, the

[36] [1897] 2 QB 57.
[37] *Wainwright*, above n 34, at [30].
[38] Ibid.
[39] Ibid, at [31].
[40] Ibid, at [18].
[41] *Wainwright v UK* [2006] ECHR 807. The court found, at [49], that the 'searches carried out on the applicants cannot be regarded as "necessary in a democratic society", within the meaning of Article 8 paragraph 2 of the Convention'.
[42] *Wainwright*, above n 34, at [18].
[43] *Campbell v MGN Ltd* [2004] UKHL 22; [2004] 2 AC 457 (HL) [11].

duty of confidence remains the predominant means for protecting informational privacy in English law.[44]

(b) Wrongs

The next question is whether existing causes of action in English law can be developed to recognise and protect a living person's right to control separated biological materials. One commentator observes that '[i]t may be that such interests [of autonomy and human dignity] should be accorded legal protection to the extent of a tortious claim for compensation for the injury involved in the invasion of personality'.[45]

English law has demonstrated a preference for extending existing wrongs, rather than creating new heads of liability. In *Wainwright*, Lord Hoffmann approved of Sir Robert Megarry VC's observation in *Malone v Metropolitan Police Commissioner*, that 'it is no function of the courts to legislate in a new field. The extension of the existing laws and principles is one thing, the creation of an altogether new right is another'.[46] Lord Nicholls described the extension of existing laws as 'the reasoned application of established common law principles, of greater or less generality, in current social conditions'.[47] Potential extensions of existing wrongs in English law are considered below.

(i) Duty of Confidence

The potential application of the duty of confidence to separated biological materials is examined here by first considering the doctrinal basis of the duty, and then outlining its elements in English law. This wrong is distinct from a contractual action for breach of confidence.[48] The jurisprudential foundation of the duty of confidence has been a matter of historical dispute and conceptual uncertainty long before the ECHR was incorporated into English law.[49] Judges and commentators have invoked various legal doctrines—including property,[50] contract,[51] and equitable principles[52]—to explain the theoretical foundation for the duty of

[44] *OBG Limited v Allan; Douglas v Hello! Limited; Mainstream Properties Limited v Young* [2007] UKHL 21; [2007] 2 WLR 920, [118] (Lord Hoffmann), [255] (Lord Nicholls). See, also: *Douglas v Hello! Ltd (No 3)* [2005] EWCA 595; [2006] QB 125, [53] (Lord Phillips MR, Clarke and Neuberger LJJ).

[45] W Cornish, *Intellectual Property: Omnipresent, Distracting, Irrelevant?* (Oxford, OUP, 2004) 22.

[46] [1979] Ch 344, 372; *Wainwright*, above n 34, at [19] (Lord Hoffmann).

[47] *In re Spectrum Plus Ltd (in liquidation)* [2005] UKHL 41; [2005] 2 AC 680 (HL) [33].

[48] There is an action for breach of a contractual duty of confidence (*Faccenda Chicken v Fowler* [1987] Ch 117, 135–38), and a source may have a cause of action for breach of a contractual duty in the event that the source consented to the taking of biological materials on the condition that any information taken was confidential.

[49] *Morrison v Moat* (1851) 9 Hare 241, 255; 68 ER 492, 498 (Turner V-C).

[50] *Butler v Board of Trade* [1971] Ch 680, 690; S Ricketson, 'Confidential Information—A New Proprietary Interest' (1977) 11 *Melbourne University Law Review* 223, 234.

[51] *Morrison*, above n 49, at 522 (Turner V–C); M Warby et al, 'Privacy and Confidentiality' in M Tugendhat and I Christie (eds), *The Law of Privacy and the Media* (Oxford, OUP, 2002) 198.

[52] *Campbell*, above n 43, at [43] (Lord Hoffmann); *Attorney-General v Guardian Newspapers Ltd (No 2)* [1990] 1 AC 109 (HL) 269 (Lord Griffiths).

confidence. Following *Campbell*, it appears that the doctrine's foundations remain in equity.[53] The task that remains for English law is to define what role autonomy and human dignity play in the ongoing development of the duty of confidence.

The nature of the action for breach of the duty of confidence received particular attention from the House of Lords in *Campbell.* This case concerned the publication in the *Daily Mirror* of information relating to Naomi Campbell's (then) drug addiction. The newspaper published a photograph and details of the treatment she was receiving from Narcotics Anonymous. Campbell argued that the newspaper committed a wrong by publishing the photograph and treatment details. In summary, the House of Lords held that A will commit a wrong in relation to B if A unjustifiably discloses to C information which A knows or ought to know is private information about B. In *Campbell*, the House of Lords held by a majority (Lord Hope, Baroness Hale and Lord Carswell) that the *Daily Mirror* had committed a wrong by publishing the photograph and treatment details.[54] In their dissenting speeches, Lord Nicholls and Lord Hoffmann suggested that, on the facts of the case, the defendants' freedom of expression outweighed the claimant's privacy interests.[55]

The significance of *Campbell* lies in the legal analysis of the duty of confidence, on which there appears to be 'no significant differences' between the majority and minority.[56] Importantly, Lord Hoffmann considered that, in respect of the unjustified publication of personal information, the breach of confidence protects an individual's autonomy and human dignity.[57] He succinctly concluded:

> Instead of the cause of action being based upon the duty of good faith applicable to confidential personal information and trade secrets alike, it focuses upon the protection of human autonomy and dignity—the right to control the dissemination of information about one's private life and the right to the esteem and respect of other people.[58]

The foundation of the right to control such information stems from a desire to protect the autonomy and human dignity of each individual.[59] There is also a legal question concerning the application of the Human Rights Act 1998 UK (HRA) in English law. Strictly speaking, ECHR rights under the HRA are binding on public

[53] *Campbell*, above n 43, at [43] (Lord Hoffmann). In *Campbell*, Lord Nicholls suggested, at [14], that the mischief addressed by the doctrine of confidence may better be described as a tort of 'misuse of private information'. The question of whether the duty of confidence has now evolved into this tort merits a more detailed examination. On the basis of the current state of English law, however, the doctrine of confidence has retained its classification as an equitable wrong. In *Campbell*, no other Law Lord classified the doctrine as a tort, and Lord Hoffmann specifically described the duty of confidence as an 'equitable action': *Campbell*, above n 43, at [43]. See, also: *Douglas v Hello (No 3)*, above n 44, at [96] (Lord Phillips MR). Burrows states that 'it cannot yet be said that breach of confidence (as a cause of action independent from breach of contract) is a tort': A Burrows, *Remedies for Torts and Breach of Contract* (Oxford, OUP, 3rd edn, 2004) 607.

[54] *Campbell*, above n 43, at [125] (Lord Hope), [158]–[60] (Baroness Hale), [169] (Lord Carswell).

[55] Ibid, at [28] (Lord Nicholls), [77] (Lord Hoffmann).

[56] Ibid, at [36] (Lord Hoffmann).

[57] Ibid, at [51] (Lord Hoffmann).

[58] Ibid.

[59] Ibid, at [50] (Lord Hoffmann).

authorities alone. Under section 6(1) of the HRA it is unlawful for a public authority (including a court) to act in a way that is incompatible with an ECHR right.[60] After *Campbell*, it appears that the court's duties under section 6 indirectly give ECHR rights horizontal effect, by virtue of the section 6 requirement that judicial decisions be compatible with ECHR rights even where proceedings exclusively involve private parties.[61] As Lord Nicholls observed in *Campbell*: 'the values underlying articles 8 and 10 are not confined to disputes between individuals and public authorities'.[62]

The more significant issue for the purposes of the current analysis is the nature of the protection afforded by the action for breach of confidence in English law. In essence, the action for breach of confidence provides a cause of action to ensure that confidential information is not used for any unauthorised purpose or disclosed without permission. The traditional formulation of the duty was set out in *Coco v AN Clark (Engineers) Ltd*,[63] which stipulated three requirements: (1) that the information have the necessary quality of confidence about it; (2) that the information be imparted in circumstances importing an obligation of confidence; and (3) that there be an unauthorised use or disclosure of the information to the detriment of the party that originally communicated it.[64] Recent judicial authority has dispensed with the second requirement. In *Attorney-General v Guardian Newspapers Ltd (No 2)*,[65] Lord Goff suggested that the duty of confidence could apply outside the context of a confidential relationship.[66] The House of Lords developed Lord Goff's suggestion in *Campbell*. Since *Campbell*, the second requirement—the existence of a confidential relationship—is no longer necessary. Lord Nicholls made this observation:

> This cause of action has now firmly shaken off the limiting constraint of the need for an initial confidential relationship. In doing so it has changed its nature . . . Now the law imposes a 'duty of confidence' whenever a person receives information he knows or ought to know is fairly and reasonably to be regarded as confidential.[67]

Following this interpretation, the duty of confidence applies whenever information is disclosed privately and in confidence, or where someone comes into possession of information that is clearly confidential.[68] The central question is, simply, whether any given information should be classified as confidential. In this way, *Campbell* represents a substantial extension of the duty of confidence.

60 HRA, s 6(3)(a).
61 *Campbell*, above n 43, at [18] (Lord Nicholls), [50] (Lord Hoffmann).
62 Ibid, at [18] (Lord Nicholls).
63 [1969] RPC 41.
64 *Coco*, ibid at 47.
65 [1990] 1 AC 109 (HL).
66 *Guardian Newspapers Ltd (No 2)*, ibid at 281.
67 *Campbell*, above n 43 at [14] (Lord Nicholls).
68 Ibid, at [14] (Lord Nicholls), [51] (Lord Hoffmann), [85] (Lord Hope), [134] (Baroness Hale), [163] (Lord Carswell).

A breach of the duty of confidence attracts three potential remedies. First, claimants may be granted an injunction to prevent disclosure.[69] Secondly, a claimant may seek damages. Although the nature and quantum of damages available are unsettled,[70] a claim can include damages for emotional distress and anxiety. In *Campbell*, for example, the House of Lords upheld an award by the trial judge of £2,500 in compensatory damages for mental distress and £1,000 in aggravated damages. It is likely that such awards will be below the level of general damages for serious physical or psychiatric injury.[71] Thirdly, a claimant may also seek delivery up of the material containing confidential information.[72]

The contentious issue is whether the concept of informational privacy (as developed in English law) can be extended to provide protection for separated biological materials. It is important for individuals to retain control over separated biological materials, not least because genetic information can be extracted from them. Genetic information is private information subject to protection, both under the duty of confidence and the DPA. It remains to be seen whether protection could be extended to embrace biological materials *before* genetic information has been extracted from them. Lord Nicholls recognised in *Campbell* that '[a]n individual's privacy can be invaded in ways not involving publication of information. Strip-searches are an example'.[73] In this way, there may be some potential for the extension of the traditional avenues for protecting information privacy interests in English law.

Two factors support an extension of the duty of confidence to cover separated biological materials themselves. First, the duty of confidence has a primary role in protecting privacy concerns in English law. For now, the House of Lords decision in *Wainwright* has effectively foreclosed debate on whether English law recognises a general right to privacy. The question that remains is: how should the concept of autonomy inform or influence the development of the duty of confidence? Whenever the legitimate privacy concerns of an individual are threatened, English law should be willing to protect them. English law adapts to changing social conditions, and separated biological materials present a novel problem requiring attention. Biological materials are unique vehicles through which the personal genetic information of an individual can be accessed. Even if such materials are simply held without an individual's consent, there remains a potential threat to that person's privacy and a threat to their autonomy and human dignity. In *R (S)*, Baroness Hale observed that: 'it is an interference with privacy for someone to know or have access to private information even if they make no use of it'.[74] Thus,

[69] *Douglas*, above n 32, at [145] (Keene LJ).

[70] For a discussion of damages in the context of the duty of confidence, see: T Aplin, 'The Development of the Action for Breach of Confidence in a Post-HRA Era' (2007) 1 *Intellectual Property Quarterly* 19, 53–59; Burrows, above n 53, at 606–11.

[71] *Archer v Williams* [2003] EWHC 1670; [2003] EMLR 38, [76] (Jackson J).

[72] *Industrial Furnaces Ltd v Reaves* [1970] RPC 605, 627 (Ch).

[73] *Campbell*, above n 43, at [15].

[74] *R (S)* HL, above n 7, at [73].

the threat to autonomy arises because of the *potential* to extract and use personal genetic information from biological materials contrary to an individual's wishes.

The issue is whether an extension of the duty of confidence to encompass physical objects would distort the concept of informational privacy. *Campbell* established that the duty of confidence protects private information. In that case Lord Nicholls also suggested that the privacy of an individual may be threatened in circumstances beyond the wrongful disclosure of information.[75] It was not necessary in *Campbell* for the House of Lords to consider how English law might protect such wider interests.

The second factor supporting the extension of the duty of confidence to cover separated biological materials is the criminal offence created under section 45 of the HTA 2004. Chapter four analysed section 45, which creates an offence where a person holds bodily materials with the intention to submit the materials for non-consensual DNA analysis.[76] Under this offence, it is not necessary for the information to be extracted and/or analysed. One of the aims in classifying the new offence is to protect the autonomy of individuals.[77] The broad ambit of the terms defining the offence indicates that Parliament considered it possible for the autonomy and human dignity of an individual to be threatened even without the extraction of personal genetic information.

These two factors indicate that the concept of informational privacy might warrant development of rights to separated biological materials. This book does not, however, support such a development because the notion of privacy, as it has developed under the duty of confidence, should be limited to the unlawful disclosure of *information*. In English law, the duty of confidence has been traditionally limited to informational privacy.[78] The duty of confidence has not provided wider protection for physical objects. In *Campbell*, the House of Lords did extend the doctrine of confidence to the publication of a photograph, thereby establishing that the publication of a photograph is analogous to the disclosure of information.[79] However, *Campbell* cannot be interpreted as extending the duty to cover physical objects in general. Although separated biological materials contain the personal genetic information of each individual source, they are not analogous to photographs. It is necessary to analyse and extract the genetic

[75] *Campbell*, above n 43 at [15].

[76] See nn 90–99 of ch 4 and accompanying text.

[77] Although the HTA 2004 Explanatory Notes make no reference to autonomy, this is a reasonable inference because the s 45 offence was drafted on the basis of a recommendation by the Human Genetics Commission (HGC), which was proposed to protect individual privacy: *Inside Information* (Department of Health, London, 2002) [3.51]. The Australian Law Reform Commission, when recommending a similar offence, observed that: 'the most obvious harm arising from testing of the sample is the intrusion on basic human dignity and autonomy': Australian Law Reform Commission, *Essentially Yours: The Protection of Human Genetic Information* (ALRC 96 Australian Government Publishing Service, Canberra, 2003) [12.2].

[78] *Campbell*, above n 43, at [51] (Lord Hoffmann).

[79] *Douglas* CA, above n 44, at [84]–[91] confirmed that publishing photographs is similar, but perhaps not identical, to publishing information. See, also: *Douglas* HL, above n 44, at [119] (Lord Hoffmann).

information from the biological materials before that information can be perceived and understood. By contrast, publication of a photograph itself—without any additional processing, analysis or extraction—presents a visual image that conveys information.

Extending the duty of confidence to encompass physical objects which themselves do not directly reveal information would blur the external boundaries of the duty. Limiting the duty of confidence to informational privacy is consistent with Lord Hoffmann's warning in *Hunter v Canary Wharf Ltd*, that 'the development of the common law should be rational and coherent. It should not distort its principles and create anomalies merely as an expedient to fill a gap'.[80] The duty of confidence can provide protection in relation to personal genetic information once it is extracted from separated biological materials. It cannot properly protect any broader right over separated biological materials.

(ii) Respect for Private and Family Life

A separate avenue exists under the HRA. In *Re Organ Retention*, Gage J considered the application of article 8(1) of the ECHR to a hypothetical matrix. In the hypothetical situation, a pathologist used the brain of a deceased child in a research project without the consent of the deceased's parents.[81] Gage J considered that the use of a child's brain in this way would constitute an infringement of article 8 of the ECHR. Further, he considered that such research was unlikely to be justifiable under article 8(2), even though he noted that each case will turn on its own facts.[82]

Although article 8 offers some protection, its application is limited to the activities of public authorities. Section 6 of the HRA provides that public authorities are required to act consistently with the ECHR. In the event that a public authority acts inconsistently with article 8, monetary compensation can be awarded under section 8 of the HRA. The definition of a public authority under the HRA is limited. Section 6(3)(a) sets out that a 'pure' public authority, such as a government department or local authority, is required to comply with ECHR rights. In *Aston Cantlow and Wilmcote with Billesley Parochial Church Council v Wallbank*,[83] the House of Lords favoured a relatively narrow test to define a 'pure' public authority. Under section 6(3)(b) of the HRA, entities that are not 'pure' public authorities, but exercise some public functions, are also required to comply with ECHR rights when exercising a 'function of a public nature'. In *Aston Cantlow*, Lord Hope noted that this second category of 'functional' authority had a much wider reach than the 'pure' public authority category.[84] Lord Nicholls observed

[80] *Hunter v Canary Wharf Ltd* [1997] AC 655 (HL) 707. See, also: *Douglas* HL, above n 44, at [99] (Lord Hoffmann).

[81] *Re Organ Retention*, above n 10, at [288]–[89].

[82] Ibid, at [298].

[83] [2003] UKHL 37; [2004] 1 AC 546 (HL).

[84] *Aston Cantlow*, above n 83, at [41]. This broad construction was noted with approval in *YL v Birmingham City Council & Ors* [2007] UKHL 27 at [4] (Lord Bingham).

that, although there was no test of universal application to determine whether a function was 'public', factors to be taken into account included:

> The extent to which in carrying out the relevant function the body is publicly funded, or is exercising statutory powers, or is taking the place of central government or local authorities, or is providing a public service.[85]

There are a number of bodies engaged in scientific research that potentially fall within the statutory meaning of public authority. NHS hospitals are public authorities.[86] More difficult questions arise in relation to the operation of genetic databases, where activities are more likely to be of a private and commercial nature. The UK Biobank is a particularly interesting example. Even though the UK Biobank receives public funding, it has no statutory basis and performs functions of a public and private nature.[87] Thus, although article 8 provides some protection, it is not clear to what range of organisations the HRA will extend.

(iii) Battery

The tort of battery concerns intentional physical contact that is not generally acceptable in the ordinary conduct of daily life.[88] Importantly, battery protects individuals not only against bodily harm, but also against interference with their right to autonomy. Three points reinforce the fact that individual autonomy is also an interest that the tort of battery seeks to protect. First, battery covers any physical contact and there is no requirement for substantial physical harm (in contrast with those torts which redress actual bodily harm). Liability attaches wherever there is unauthorised contact which is not 'generally acceptable in the ordinary conduct of daily life'.[89] Secondly, US judicial authority[90] and academic commentary[91] support the application of battery even where a claimant is unaware of the occurrence of unlawful contact. The claimant need not be aware of the physical contact because the law considers the mere violation of an individual's autonomy to be sufficient. Thirdly, the centrality of autonomy emerged after English law considered whether the tort of battery required the mere intention to commit the

[85] *Aston Cantlow*, above n 83, at [12]. See, also: *Birmingham City Council*, above n 84, at [5]–[13] (Lord Bingham), [64] (Baroness Hale), [91] (Lord Mance). For an academic discussion of these factors, see: R Clayton, 'The Human Rights Act Six Years On: Where Are We Now?' [2007] 1 *European Human Rights Law Review* 11, 14–16 and H Quane, 'The Strasbourg Jurisprudence and the Meaning of a "Public Authority" Under the Human Rights Act' [2006] *Public Law* 106.

[86] *Re Organ Retention*, above n 10.

[87] In *Birmingham City Council*, above n 84, Lord Bingham noted at [8] 'that the absence of any statutory intervention will tend to indicate parliamentary recognition that the function in question is private and so an inappropriate subject for public regulation'. For an academic discussion of the UK Biobank, see: C Johnston and J Kaye, 'Does the UK Biobank Have a Legal Obligation to Feedback Individual Findings to Participants?' (2004) 12 *Medical Law Review* 239, 263.

[88] See nn 119–26 of ch 1 and accompanying text.

[89] *Collins v Wilcock* [1984] 1 WLR 1172 (CA) 1177 (Robert Goff LJ).

[90] *Restatement of the Law, Second, Torts*, § 18, comment (d). See, also: *Hively v Higgs* 253 P 363 (Or SC 1927) 365; *Mohr v Williams* 104 NW 12 (Minn SC 1905) 271.

[91] J Fleming, *The Law of Torts* (Sydney, LBC, 9th edn, 2002) 30.

act (of touching), rather than an intention to cause harm. In *Wilson v Pringle*, Croom-Johnson LJ held that an 'intention to injure is not essential to an action for trespass to the person. It is the mere trespass by itself which is the offence'.[92] This focus supports the proposition that the tort of battery seeks to protect individual autonomy.

The pertinent issue here is whether the right to bodily integrity (as protected by the tort of battery) can be extended to include the protection of separated biological materials. Again, even if it is conceptually possible, such an extension would disfigure the tort of battery as it has developed in English law. The tort has evolved so as to protect invasions of a person's body. The extension of the tort of battery to separated biological materials would require that those materials remain somehow equivalent to the person. This is a significant and unsustainable conceptual leap. Separated biological materials are distinct from the person from whom they originated and the tort of battery should be limited to protecting the person as a whole.

(iv) Negligence

A third potential avenue for safeguarding an individual's non-proprietary interests in biological materials lies with an award of compensation for any loss of autonomy suffered under the tort of negligence. Traditionally, negligence has been limited to negligently inflicted damage to an individual's property or invasions of bodily integrity. In *Rees v Darlington Memorial Hospital NHS Trust*,[93] the House of Lords recognised loss of autonomy as a head of damage in the context of a negligence claim. There, the claimant suffered from a severe visual handicap. She gave birth to a healthy son, having previously undergone a sterilisation operation that was performed negligently in a hospital managed by the defendants. The claimant brought an action claiming damages for the costs involved in raising her son. At first instance, the judge held that *McFarlane v Tayside Health Board*[94] prevented recovery of such costs. In *McFarlane*, the House of Lords considered the extent to which damages are recoverable for the birth of an unintended child following a wrongful pregnancy. The claimant became pregnant following a failed sterilisation and she brought an action claiming the costs of bringing up a normal, healthy child. In *McFarlane*, the House of Lords unanimously held that the costs of raising a child were not recoverable.[95] In *Rees*, a majority of the Court of Appeal (Hale and Robert Walker LJJ, Waller LJ dissenting) held that it was fair, just and reasonable for a disabled parent to claim the additional costs of raising a child, which could be attributed to her disability.[96] The defendant appealed.

[92] [1987] 1 QB 237, 249.

[93] [2003] UKHL 52; [2004] 1 AC 309 (HL).

[94] [2000] 2 AC 59 (HL).

[95] *McFarlane*, ibid at 76 (Lord Slynn), 83 (Lord Steyn), 97 (Lord Hope), 106 (Lord Clyde), 113–14 (Lord Millett).

[96] *Rees v Darlington Memorial Hospital NHS Trust* [2002] EWCA Civ 88; [2003] QB 20, [23]–[25] (Hale LJ), [37] (Robert Walker LJ), [55] (Waller LJ).

A majority of the House of Lords (Lords Bingham, Nicholls, Millett and Scott; Lords Steyn, Hope and Hutton dissenting) allowed the appeal. The majority held that the claimant could not recover damages for the cost of raising a normal healthy child. However, they did hold that the claimant was entitled to an award of damages for the pain and costs associated with the pregnancy and birth itself (as in *McFarlane*), as well as an award of £15,000 for her loss of autonomy.[97] The majority's award of £15,000 for the loss of autonomy was drawn from Lord Millett's speech in *McFarlane*, where he held that the claimant was entitled to an award of £5,000 for the denial of an aspect of her autonomy.[98] In *McFarlane*, Lord Millett observed that the parents had 'lost the freedom to limit the size of their family. They have been denied an important aspect of their autonomy'.[99]

Forceful objections can be directed against the decision in *Rees* to award damages for a loss of autonomy. Why should the compensation be fixed at £15,000? This award is an insufficient and arbitrary response to the invasion of a woman's reproductive autonomy. The most interesting aspect of the decision, for present purposes, is the majority's justification for the award of general damages. Lord Nicholls and Lord Millett were concerned with protecting the claimant's loss of autonomy. Lord Nicholls concluded:

> An award of some amount should be made to recognise that in respect of birth of the child the parent has suffered a legal wrong, a legal wrong having a far-reaching effect on the lives of the parent and any family she may already have.[100]

On the same point, Lord Millett observed:

> I still regard the proper outcome in all these cases is to award the parents a modest conventional sum by way of general damages, not for the birth of the child, but for the denial of an important aspect of their personal autonomy, viz the right to limit the size of their family. This is an important aspect of human dignity, which is increasingly being regarded as an important human right which should be protected by law. The loss of this right is not an abstract or theoretical one. As my noble and learned friend Lord Bingham of Cornhill has pointed out, the parents have lost the opportunity to live their lives in the way that they wished and planned to do. The loss of this opportunity, whether characterised as a right or a freedom, is a proper subject for compensation by way of damages.[101]

Lord Bingham stated that the loss suffered by the claimant amounted to a denial of the 'opportunity to live her life the way that she wished and planned'.[102] Although this statement suggests that Lord Bingham was also concerned with the loss of autonomy, he commented that the conventional award was not compensatory.[103] It

[97] *Rees*, above n 93, at [8] (Lord Bingham), [17] (Lord Nicholls), [125] (Lord Millett), [148] (Lord Scott).

[98] *McFarlane*, above n 94, at 114.

[99] Ibid.

[100] *Rees*, above n 93, at [17].

[101] Ibid, at [123].

[102] Ibid, at [8].

[103] Ibid.

is difficult to understand the basis of the award, however, if it is not compensatory. The fourth member of the majority, Lord Scott, awarded damages on the basis that the claimant was 'deprived of the benefit she was entitled to expect'.[104]

Lord Nicholls and Lord Millett, and perhaps Lord Bingham, considered that the award remedied the loss of autonomy.[105] On this view, the award of £15,000 can be viewed as non-pecuniary damages flowing from the negligent sterilisation procedure. The negligence resulted in an interference with the claimant's autonomy—an infringement of an interest protected by law—and required compensation.

In the event that a claimant could establish a negligence claim, English law supports a head of damage for the loss of autonomy. At present, such damages have only been awarded in the narrow context of a negligence claim for wrongful birth, but the possibility of further developing the law of negligence to include non-consensual use of separated biological materials warrants closer analysis. To bring a claim in negligence for non-consensual use, a claimant would have to establish the three main elements: (1) the existence of a duty of care; (2) a breach of that duty as a consequence of the defendant's failure to take reasonable care; and (3) a sufficient causative link between the breach and the loss of autonomy sustained by the claimant.[106]

English law recognises that damages will only be awarded if the loss that the claimant suffers falls within the scope of the defendant's duty of care.[107] Establishing a duty of care may be the principal obstacle in the context of protecting an individual's right to autonomy over separated biological materials. Even if a duty can be established, the claimant must show also that the defendant committed a negligent act. It may be that the defendant's action did not fall below the requisite standard of care.

The factual scenario developed in Part II, concerning the removal of biological materials (X) from a source (A) for utilisation in research, can also assist in analysing non-proprietary protection.[108] In terms of negligence liability, the question is whether a duty of care falls on either the person who removes the biological materials (B), the researcher who uses them (C), or the employer of B and C, such as a hospital (D). *Re Organ Retention Group Litigation*[109] represents a recent extension of a duty of care in the context of separated bodily materials from corpses. In *Re Organ Retention*, Gage J found that the doctor responsible for the post-mortem examination owed a duty of care to the parents of the deceased child.[110] If English

[104] *Rees*, above n 93, at [148].

[105] For a different view, see: Burrows, above n 53, at 291. Burrows argues that, based on the view of Lord Scott, the award of damages represented compensation for the claimant's mental distress at having her lifestyle plans disrupted. It seems, however, that based on the speeches of Lord Nicholls and Lord Millett, a more compelling interpretation is that the compensation represented an award for the loss of autonomy.

[106] Davies, above n 14, at 415.

[107] *Chester v Afshar* [2004] UKHL 41; [2005] 1 AC 134 (HL) [51] (Lord Hope).

[108] See n 22 of ch 5 and accompanying text.

[109] *Re Organ Retention*, above n 10.

[110] Ibid, at [200].

law is willing to recognise a duty of care in the context of materials removed from a deceased child, then, there seems to be no basis in principle for objecting to the extension of this duty to materials removed from a living person.

Biological materials will be removed from living persons in many different circumstances. In some cases, they will be removed in the course of medical treatment. In other cases, they will be removed specifically for research purposes, such as for inclusion in a genetic database. Nevertheless, where biological materials are collected for the specific purpose of medical research, such as in the case of UK Biobank, it is arguable that a duty of care exists.[111] The scope of the duty of care would include, for example, disclosing to the patient the purposes for which separated biological materials are intended or likely to be used.

It is not suggested that the law of negligence should be used as the main vehicle in English law to protect an individual's non-proprietary interests in separated biological materials. There may be instances where a negligence claim cannot be established because no duty of care exists, or, if it does, there has been no breach of that duty. Further, a more principled and coherent approach lies in the creation of an independent tort that would protect an individual's distinct autonomy interest in separated biological materials. Before turning to the potential development of a new liability, it is necessary to also consider whether English law can protect an individual's right to be free from emotional distress.

2 Emotional Distress

English law provides protection for the negligent infliction of a psychiatric illness.[112] In *Re Organ Retention*, Gage J made an award for psychiatric damages suffered by one of the claimants.[113] However, the difficulties in establishing such a claim in negligence were demonstrated by the fact that similar claims were dismissed.[114] In many cases it is unlikely that a living person will develop a psychiatric illness following the non-consensual use of their biological materials. It is plausible that an individual (or relative of a deceased person) may suffer emotional distress. Freedom from emotional distress is the second potential interest that a living person may assert over the non-consensual use of separated biological materials. Relatives of a deceased person can also assert such an interest over the non-consensual use of separated biological materials from that person. The question, then, is whether English law can provide protection from such emotional distress.

[111] Johnston and Kaye, above n 87, at 252.

[112] *McLoughlin v O'Brian* [1983] 1 AC 410 (HL) 431 (Lord Bridge). See, also: *Alcock v Chief Constable of South Yorkshire Police* [1992] 1 AC 310 (HL) 401 (Lord Ackner).

[113] *Re Organ Retention*, above n 10, at [267].

[114] In Harris's case, Gage J held that the psychiatric injury was not reasonably foreseeable: *Re Organ Retention*, ibid at [259]. In Carpenter's case, Gage J held that the knowledge of organ retention did not materially contribute to Carpenter's psychiatric illness: *Re Organ Retention*, ibid at [278].

(a) Definition

Although the distinction between emotional (or mental) distress and psychiatric injury is often not clear as a matter of medicine,[115] English law draws a distinction.[116] Smith has defined emotional distress as being any 'disagreeable disturbance of emotional or mental tranquillity'.[117] Emotional distress is defined broadly here to include distress, suffering and grief, injury to pride or feelings, and diminution in self-respect, which does not amount to a psychiatric illness.

(b) Wrongs

English law awards damages for emotional distress if it is inflicted in the course of another wrong, such as trespass,[118] defamation[119] and fraud.[120] In so doing, English law acknowledges that emotional distress can be a source of loss or damage. English law does not currently recognise the infliction of emotional distress as a separate cause of action. There are three avenues by which English law could provide a cause of action where a claimant suffers emotional distress through the non-consensual use of separated biological materials: (1) the *Wilkinson v Downton*[121] principle; (2) harassment; and (3) intentional infliction of emotional distress.

(i) Wilkinson v Downton

Wilkinson v Downton is often referred to by commentators as having established a tort of intentionally or recklessly causing harm to an individual amounting to psychiatric damage.[122] *Wilkinson* concerned the actions of Mr Wilkinson, who went to the races by train and left his wife at home, and Downton, who played a practical joke on Mrs Wilkinson by telling her that Mr Wilkinson had been seriously injured in an accident. The story was false and Mr Wilkinson returned that evening. However, the story had a dramatic impact on Mrs Wilkinson and she fell ill. A claim in negligence was likely to fail because her nervous shock was too remote.[123] Wright J found, nevertheless, that a cause of action lay:

> The defendant has, I assume for the moment, wilfully done an act calculated to cause physical harm to the plaintiff—that is to say, to infringe her legal right to physical safety, and has in fact caused physical harm to her. That proposition without more appears to me to state a good cause of action, there being no justification alleged for the act. This

[115] Law Commission, *Liability for Psychiatric Illness* (Law Commission Report 249, London, 1998) 51.

[116] *Page v Smith* [1996] 1 AC 155 (HL) 167 (Lord Keith).

[117] HW Smith, 'Problems of Proof in Psychiatric Injury Cases' (1963) 14 *Syracuse Law Review* 586, 610.

[118] *Drane v Evangelou* [1978] 1 WLR 455 (CA) 459.

[119] *Cassell & Co v Broome* [1972] AC 1027 (HL) 1085.

[120] *Archer v Brown* [1985] QB 401, 426.

[121] [1897] 2 QB 57.

[122] PR Handford, *Mullany and Handford's Tort Liability for Psychiatric Damage* (Sydney, Lawbook, 2nd edn, 2006) 678; R Townshend-Smith, 'Harassment as a Tort in English and American Law: The Boundaries of *Wilkinson v Downton*' (1995) 24 *Anglo-American Law Review* 299, 325–26.

[123] *Wilkinson v Downton* [1897] 2 QB 57, 59–60.

> wilful injuria is in law malicious, although no malicious purpose to cause the harm which was caused nor any motive of spite is imputed to the defendant.[124]

It is not clear on what basis Wright J extended liability. One interpretation of Wright J's decision is that he recognised a new tort of intentionally or recklessly causing harm to an individual amounting to psychiatric damage.[125] With the exception of the US,[126] *Wilkinson* had not, until recently, been the subject of extensive judicial consideration. The paucity of case law in Commonwealth countries has changed over the last fifteen years as courts in England,[127] Australia[128] and Canada[129] have considered claims based on the *Wilkinson* principle. English[130] and Australian[131] law both appear to have rejected the existence of a separate *Wilkinson* tort, while Canadian law has embraced it.[132]

The scope of the *Wilkinson* principle was considered by the House of Lords in *Wainwright v Home Office.*[133] In *Wainwright*, the claimants submitted that the unauthorised strip-search infringed the principle established by *Wilkinson.* As already noted, Lord Hoffmann delivered the leading speech, and held that the principle in *Wilkinson* was now effectively part of the law of negligence.[134] Lord Scott reached a similar conclusion.[135] Lord Hoffmann noted that Wright J's decision should be viewed as an attempt to circumvent *Victorian Railways Commissioners v Coultas*,[136] where the Privy Council held that nervous shock was too remote a consequence of a negligent act to be a recoverable head of damage.[137] Seen in this light, Wright J's imputation of an intention to harm on the part of the defendant was important to distinguish the Privy Council authority.

124 Ibid, at 59.

125 PR Glazebrook, '*Wilkinson v Downton*: A Centenary Postscript' (1997) 32 *Irish Jurist* 46, 48; PR Handford, '*Wilkinson v Downton* and Acts Calculated to Cause Physical Harm' (1985) 16 *University of Western Australia Law Review* 31, 33.

126 US jurisprudence has long recognised a cause of action on similar terms to the principle in *Wilkinson*. See, for example: *Voss v Bolzenius* 128 SW 1 (Miss CA 1910); *Goddard v Watters* 82 SE 304 (Geo CA 1914); *Johnson v Sampson* 208 NW 814 (Minn SC 1926); *Great Atlantic & Pacific Tea Co v Roch* 153 A 22 (Md CA 1930). For a discussion of the development of this tort in the US, see: WL Prosser, 'Intentional Infliction of Mental Suffering—A New Tort' (1939) 37 *Michigan Law Review* 874.

127 *Wainwright*, above n 34; *Wong v Parkside Health NHS Trust* [2001] EWCA Civ 1721; [2003] 3 All ER 932 (CA); *Wainwright v Home Office* [2001] EWCA Civ 2081; [2002] QB 1334; *Powell v Boladz* [1998] Lloyd's Rep Med 116 (CA); *Khorasandjian v Bush* [1993] 1 QB 727; *Burnett v George* [1992] 1 FLR 525 (CA).

128 *Carrier v Bonham* [2002] 1 Qd R 474 (Qld CA); *Coyne v Commercial Equity Corporation Ltd* (1998) 20 WAR 109 (WA SC); *Gimson v Victoria Workcover Authority* [1995] 1 VR 209 (Vic SC); *Pavlovic v Commonwealth Bank of Australia* (1991) 56 SASR 587 (SA SC).

129 *Clark v Canada* (1994) 20 CCLT 241 (FCTD); *Boothman v Canada* [1993] 3 FC 381; *Timmermans v Buelow* (1984) 38 CCLT 136 (FCTD); *Rahemtulla v Vanfed Credit Union* (1984) 29 CCLT 78 (BC SC).

130 *Wainwright*, above n 34.

131 *Carrier*, above n 128, at 484; See, also: *Magill v Magill* [2006] HCA 51; (2006) 231 ALR 277 (HCA) [20] (Gleeson CJ), [117] (Gummow, Kirby and Crennan JJ).

132 *Clark*, above n 129.

133 *Wainwright*, above n 34.

134 Ibid, at [47].

135 Ibid, at [62].

136 (1888) 13 App Cas 222.

137 *Coultas*, ibid at 225.

After *Wilkinson*, nervous shock became actionable in the English law of negligence early in the twentieth century.[138] Indeed, in *Wainwright* Lord Hoffmann concluded that the principle in *Wilkinson* had been subsumed into the law on the negligent infliction of psychiatric injury.[139] As noted above, it is unlikely that an individual will develop a psychiatric illness following the non-consensual use of biological materials.

Further, in *Wainwright* Lord Hoffmann considered that if English law were to recognise liability for the intentional infliction of emotional distress, imputed intention would not be sufficient. For Lord Hoffmann, the defendant must 'have intended to cause harm or at least acted without caring whether he caused harm or not'.[140] On the facts in *Wainwright*, he considered that in any event, the necessary intention was not established.[141] The protection against emotional distress in English law must therefore be located elsewhere.

(ii) Harassment

In addition to *Wilkinson*, there have been judicial suggestions that English law recognises a tort of harassment. The potential for such a tort arose in *Khorasandjian v Bush*,[142] which concerned the breakdown of a relationship and subsequent pestering of the plaintiff through threats of violence and persistent telephone calls by the defendant. In *Khorasandjian*, the Court of Appeal (Dillon, Rose LJJ; Peter Gibson LJ dissenting) made two main points. First, it held that harassment through telephone calls interfering with an individual's enjoyment of property could be restrained by a *quia timet* injunction as a private nuisance.[143] Secondly, the Court of Appeal held that harassment not amounting to threats but likely to cause physical or psychiatric illness also may be restrained by a *quia timet* injunction based on the principle in *Wilkinson*.[144]

If the Court of Appeal did in fact (implicitly) recognise a tort of harassment in *Khorasandjian*, the tort would seem to protect the autonomy and human dignity of an individual. A more likely interpretation of *Khorasandjian* is that the decision simply represents an extension of the tort of private nuisance. Indeed, the House of Lords held in a later case—*Hunter v Canary Wharf Ltd*[145]—that the Court of Appeal overstretched the tort of nuisance in *Khorasandjian*. In *Canary Wharf*, Lord Goff ruled that the Court of Appeal had used the law of nuisance to create a tort of harassment via the 'back door'.[146] The Court of Appeal subsequently

138 *Dulieu v White & Sons* [1901] 2 KB 669, 672.
139 *Wainwright*, above n 34, at [41], [47].
140 Ibid, at [45].
141 Ibid.
142 [1993] 1 QB 727.
143 Ibid, at 734–35 (Dillon LJ).
144 Ibid, at 736 (Dillon LJ).
145 Above n 80.
146 *Canary Wharf*, ibid at 691–92.

acknowledged that English law had not developed a tort of harassment in *Wong v Parkside Health NHS Trust.*[147] In that case, Hale LJ noted:

> The case law was developing in such a way that some considered that a new tort of harassment had already been created. However, in all of the cases the question was whether an injunction should be granted to prevent particular kinds of behaviour: there was no case in which damages were awarded to compensate for such behaviour in the past.[148]

The potential development of a tort of harassment was nevertheless superseded by the introduction of a new statutory tort providing a right of action for harassment under the Protection from Harassment Act 1997 UK (Harassment Act). As Lord Hoffmann observed in *Canary Wharf*: '[t]he law of harassment has now been put on a statutory basis . . . and it is unnecessary to consider how the common law might have developed'.[149] Lord Goff made a similar point.[150]

The Harassment Act, at its core, introduced criminal and civil liability for any 'course of conduct which amounts to harassment of another'.[151] The Act provides damages for anxiety and was introduced to redress the inadequacy of the law concerning the infliction of distress and humiliation by stalking and other forms of pestering.[152] At present, the Harassment Act offers the only form of protection against the infliction of emotional distress under English law.[153] The Act cannot, however, offer protection for the non-consensual use of separated biological materials because it is unlikely that such action could constitute harassment or be classified as a course of conduct within the terms of the Harassment Act.

(iii) Intentional Infliction of Emotional Distress

Although a tort of intentional infliction of emotional distress is not currently part of English law, Lord Hoffmann has indicated that English law could develop a remedy, and he noted in *Hunter v Canary Wharf Ltd* that there is 'no reason why a tort of intention should be subject to the rule which excludes compensation for mere distress, inconvenience, or discomfort in actions based on negligence . . . The policy considerations are quite different'.[154] More recently, in *Wainwright*, Lord Hoffmann commented:

> I do not resile from the proposition that the policy considerations which limit the heads of recoverable damage in negligence do not apply equally to torts of intention. If someone actually intends to cause harm by a wrongful act and does so, there is ordinarily no reason why he should not have to pay compensation. But I think that if you adopt such a principle, you have to be very careful about what you mean by intention . . . Even on

[147] [2001] EWCA Civ 1721; [2003] 3 All ER 932 (CA).
[148] *Wong*, ibid at 941.
[149] *Canary Wharf*, above n 80, at 707.
[150] Ibid, at 691–92.
[151] Harassment Act, s 1.
[152] Harassment Act, s 3(2).
[153] See, for example: *Potter v Price* [2004] EWHC 781.
[154] *Canary Wharf*, above n 80, at 707.

> the basis of a genuine intention to cause distress, I would wish, as in *Hunter's Case* [1997] AC 665, to reserve my opinion on whether compensation should be recoverable.[155]

In neither *Canary Wharf* nor *Wainwright* did Lord Hoffmann consider the potential development of such a tort in any detail. Nevertheless, for an individual to be protected from the infliction of emotional distress, it is necessary that English law develop a tort as suggested by Lord Hoffmann.

Other common law jurisdictions have adopted different positions. No such tort is recognised in Australian law.[156] Support for a tort along these lines can be found in other jurisdictions including, the US,[157] and Canada.[158] In the US, for example, the *Restatement of the Law, Torts* was amended so as to recognise liability for the intentional infliction of emotional distress.[159] There is now a substantial body of American jurisprudence addressing this point, and one commentator has argued that every US state recognises this tort in one form or another.[160] An intentional tort operates in American case law as a 'gap-filler' for cases where traditional actions do not provide a remedy for emotional distress.[161] The US tort has three elements. First, the defendant must commit an act with an intention to cause emotional distress. It is sufficient if, instead of having the requisite intention, the defendant is reckless and demonstrates a disregard of the high probability that the conduct will cause emotional distress.[162] Secondly, the defendant's conduct must be so outrageous in character that it goes beyond all bounds of decency and must be regarded as intolerable in a civilised society.[163] Thirdly, the plaintiff must suffer emotional distress. It is difficult to determine the threshold level of emotional distress; in general, US courts have required that the infliction of emotional distress be 'so severe that no reasonable person could be expected to endure it'.[164]

Significantly, US courts have applied the tort of intentional infliction of emotional distress in the context of interference with dead bodies. Chapter two noted that US law provides a cause of action for the next-of-kin where someone intentionally inflicts emotional distress by interfering with a dead body.[165] American jurisprudence, therefore, presents a model for the development in English law of

[155] *Wainwright*, above n 34, at [44]–[46].

[156] *Giller v Procopets* [2004] VSC 113 (Vic SC) [186] (Gillard J).

[157] *Banks v Fritsch* 39 SW 3d (Ky CA 2001) 481.

[158] *Tran v Financial Recovery Debt Recovery Ltd* (2000) 193 DLR (4th) 168.

[159] *Restatement of the Law, Torts*, Supplement, § 46 (1948). The *Restatement of the Law, Second, Torts* (1965) approved this amendment and provided: '[o]ne who by extreme and outrageous conduct intentionally or recklessly causes severe emotional distress to another is subject to liability for such emotional distress and if bodily harm results from it, for such bodily harm'.

[160] ML Carpenter, '*Petersen v Sioux Valley*: Reckless Infliction of Emotional Distress' (1993) 38 *South Dakota Law Review* 359, 365.

[161] *Banks*, above n 157, at 481.

[162] *Hinkley v Baker* 122 F Supp 2d 57 (US DC 2000) 60.

[163] *Restatement of the Law, Second, Torts*, § 46, comment (d).

[164] *Restatement of the Law, Second, Torts*, § 46, comment (j). See, also: *Folz v State* 797 P 2d 246 (NW SC 1990) 254.

[165] See, for example: *Jaynes v Strong–Thorne Mortuary Inc* 954 P 2d 45 (NM SC 1997) 50; *Christensen v Superior Court* 820 P 2d 181 (Cal SC 1990) 202.

an intentional tort, and demonstrates its application in the context of interference with dead bodies.

The desirability of a tort of intentional infliction of emotional distress has also been considered recently in Scotland. In a published report, the Scottish Law Commission recommended that Scots law should create liability where a wrongdoer acts with the deliberate intention of inflicting mental injury to the victim.[166] The Commission concluded:

> [W]e have become satisfied that a distinction should be drawn between the situation where mental harm is caused deliberately by the defender and where it has arisen as a result of his unintentional but wrongful conduct, for example negligence. In the case of intentional wrongdoing, we now think that the defender should normally be liable for the harm he intended to cause: this should include distress, anxiety, grief, anger etc, whether or not this amounts to a medically recognised mental disorder.[167]

Under this approach, liability could attach for the deliberate infliction of mere emotional distress without the need for the development of a psychiatric illness. The Commission also addressed the difficult question of what form or quality of emotional distress should suffice to create liability. It concluded that the requirement would be satisfied by 'any harm to a person's mental state, mental functioning or well-being whether or not the harm amounts to a medically recognized mental disorder'.[168]

There are two main objections to a cause of action based on the infliction of emotional distress. First, there is the standard 'floodgates' argument; namely that the traditional requirement in negligence law of psychiatric harm is required to safeguard against trivial or illegitimate claims.[169] The American experience demonstrates that these fears may be misguided because control mechanisms can prevent trivial claims succeeding.[170] In the US, the 'outrageous conduct' requirement acts as the primary control mechanism.

A second objection is that emotional distress is too difficult to assess and cannot provide a clearly defined basis for liability.[171] Emotional distress is often short-lived in comparison to a psychiatric illness, making it more difficult to quantify. This objection is not sufficient to preclude recognition of liability. Temporary emotional imbalance can be intense and highly damaging to an individual.[172] Further, the difficulty of quantifying such loss has not inhibited English law from awarding damages for emotional distress in several contexts. English law

166 Scottish Law Commission, *Report on Damages for Psychiatric Injury* (Scottish Law Commission Report No 196, Edinburgh, 2004) [1.8].
167 Ibid, at [3.7].
168 Ibid, at [3.8].
169 *Giller*, above n 156, at [186] (Gillard J).
170 Handford, above n 122, at 701–02.
171 Burrows, above n 53, at 338.
172 J Yiend and T Dalgleish, 'Grief, Stress and Post-traumatic Stress Disorder' in A Warrell et al (eds), *Oxford Textbook of Medicine* (Oxford, OUP, 4th edn, 2004) 1289; D Goldberg, 'A Classification of Psychological Distress for Use in Primary Care Settings' (1992) 35 *Social Science and Medicine* 189, 192.

recognises that an individual can recover compensation for enduring an unsatisfactory holiday.[173] As noted above, English law also provides compensation for emotional distress suffered in the course of other torts, including trespass[174] and defamation.[175] Further, in a statutory context, the Harassment Act specifically provides damages for anxiety.[176]

Thus, English law can, at least in principle, develop a tort for the intentional infliction of emotional distress. Drawing on Lord Hoffmann's comments in *Wainwright*, an intentional infliction of emotional distress tort would have two essential elements: (1) the defendant commits an act which is intended to cause emotional distress; and (2) the claimant suffers emotional distress as a consequence of that act. With respect to (2), in *Wainwright* Lord Hoffmann was not clear as to what degree of distress would be required for a potential tort to attach. Although establishing a requisite standard is not straightforward, US jurisprudence suggests that an appropriate standard may be that of 'severe' emotional distress, to preclude recovery for transient distress.

The factual scenario employed throughout Part II concerns a case where A's separated biological materials (X) are used either for the creation of a cell-line without A's consent, or in research not approved by A. If either event occurs, it is conceivable that A could suffer emotional distress. But could a tort of intentional infliction of emotional distress offer A any meaningful protection? Closer analysis suggests that it would not.

The first element of the projected tort would require that the person who uses X form an intention to inflict emotional distress on A. An intention to perform the non-consensual act (research or cell-line creation) is not sufficient. In most of the practical situations considered here, researchers use separated biological materials for medical and scientific research. It is highly unlikely that they would have the necessary intention to inflict emotional distress. The second element of the proposed tort would require that A suffer severe emotional distress. Again, it may be difficult to establish severe emotional distress (even though this standard is more easily met than the requirement of a psychiatric illness under the law of negligence).

Even assuming that these elements can be satisfied, the question remains: what compensation should be available for a tort of intentional infliction of emotional distress? Lord Hoffmann deliberately left this question open in *Wainwright*,[177] and there is little guidance in English law as to what principles should guide the quantum of such damages. It is useful, therefore, to consider analogous damages awards. In the context of an action for breach of confidence, damages have been awarded for injury to feelings. In *Archer v Williams*, Jackson J observed that such awards are likely to be modest: '[g]eneral damages for injury to feelings should be kept to a modest level and should be proportionate to the injury suffered. Such

[173] *Jarvis v Swans Tours* [1973] QB 233.
[174] *Drane v Evangelou* [1978] 1 WLR 455 (CA) 459.
[175] *Cassell & Co v Broome* [1972] AC 1027 (HL) 1085.
[176] Harassment Act, s 3(2).
[177] *Wainwright*, above n 34, at [45].

awards should be well below the level of general damages for serious physical or psychiatric injury'.[178] In the event that the two elements of the suggested tort of intentional infliction of emotional distress could be established, then by analogy it is likely that English law would provide at most a modest award of damages. English law may also provide a remedy for the suggested tort by way of an injunction to prevent the future infliction of emotional distress.[179]

3 Summary

There are two interests that English law could, and arguably should, protect in relation to separated biological materials: (1) autonomy; and (2) freedom from emotional distress. Only article 8 of the ECHR and the law of negligence have the potential to provide protection in relation to autonomy. For emotional distress, even a tort of intentional infliction of emotional distress is unlikely to offer substantial protection for the misuse of separated biological materials. Thus, the current limitations of English law necessitate the creation of additional liability.

C Developing Liability

English common law is not readily open to the recognition of new forms of liability because the extension of liability is generally considered a matter for Parliament alone.[180] Statutory intervention is unlikely given the opportunity that was by-passed through the HTA 2004. The Australian Federal Parliament has shown a similar reluctance to intervene.[181] There is at present no recognisable tort of non-consensual use of separated biological materials in the common law jurisdictions of England, Australia or the US. Part II noted Scottish authority[182] concerning the availability of compensation following the wrongful interference with a cadaver. *Stevens v Yorkhill NHS Trust and Southern General University Hospitals NHS Trust* was based, however, on the existence of a Scottish *action iniuriarum* (solatium for affront). There is no equivalent right to compensation in England and, even in Scotland, the existence of the solatium for affront action is a matter of contention.[183]

178 [2003] EWHC 1670; [2003] EMLR 38, [76] (Jackson J).

179 A Burrows, 'Judicial Remedies' in P Birks (ed), *English Private Law* (Oxford, OUP, 2000) 877–79.

180 *Wainwright*, above n 34, at [33] (Lord Hoffmann); *Malone v Metropolitan Police Commissioner* [1979] Ch 344, 372 (Megarry VC).

181 Government Response to Recommendations of Australian Law Reform Commission, *Essentially Yours: The Protection of Human Genetic Information* (Commonwealth Government, Canberra, 2005).

182 *Stevens v Yorkhill NHS Trust and Southern General University Hospitals NHS Trust* [2006] CSOH 143, 2006 SLT 889 (Outer House); *Pollok v Workman* (1900) 2 F 354.

183 JK Mason, RA McCall-Smith and GT Laurie, *Law and Medical Ethics* (London, Butterworths, 6th edn, 2002) 437.

The potential for a separate common law tort of wrongful interference with separated biological materials was raised in chapter two. In *Re Organ Retention*, the claimants argued that there should be a tort of wrongful interference with cadavers in English law, although they based the existence of this wrong on the law of conversion and, hence, on alleged property rights.[184] This chapter is concerned, instead, with the protection of non-proprietary interests in separated biological materials. The suggested tortious liability requires consideration of the: (1) protected interests; (2) conduct which should attract liability; and (3) elements of the proposed liability.

1 Protected Interests

Autonomy and freedom from emotional distress have already been identified as the two non-proprietary interests that individuals have over separated biological materials. Relatives of a deceased person also have a right to freedom from emotional distress. Although an individual may suffer emotional distress following the non-consensual use of biological materials, it is more appropriate for the infliction of emotional distress to be protected through the development of a tort of general application. Such a general tort of intentional infliction of emotional distress has already been discussed.

Autonomy is the other non-proprietary interest. English law has recognised a living person's continued interests in separated biological materials. Specifically, the HTA 2004 has established a statutory regime that, in broad terms, requires persons who wish to use separated biological materials to obtain the consent of the individual from whom those materials were removed. This raises a further question: how can English law develop liability to better protect this continued interest for a living person?

2 Conduct

The answer to this question lies, in part, in determining the conduct to which liability should attach. The spectrum of conduct includes: (1) intentional conduct; (2) reckless conduct; (3) negligent conduct; and (4) accidental conduct. A tort that encompassed all four would be classified in English law as a strict liability tort. The common law is generally reluctant to impose a strict liability rule and, as Lord Goff observed in *Cambridge Water Co v Eastern Countries Leather plc*,[185] as a general principle, it is more appropriate for Parliament rather than the courts to impose new forms of strict liability.[186]

[184] *Re Organ Retention*, above n 10, at [152].
[185] [1994] 2 AC 264 (HL).
[186] *Cambridge Water*, ibid at 305.

In the context of the non-consensual use of biological materials, sufficient protection can be provided to individuals without the imposition of a strict liability rule. This conclusion is consistent with the approach adopted by the UK Parliament through the HTA 2004. Even though the HTA 2004 provides for the prosecution of criminal offences, it offers defences where a person 'reasonably believes' either that appropriate consent has been obtained for the activity being performed, or that obtaining such consent is not necessary.[187] Further, it is not necessary for liability to encompass negligent conduct. Negligent conduct can be adequately addressed through the tort of negligence.[188]

Liability under the suggested tort should, therefore, be limited to the intentional or reckless non-consensual use of separated biological materials. Reckless conduct is included in the suggested tortious liability on the basis that 'reckless indifference to consequences is as blameworthy as deliberately seeking such consequences'.[189] In tort, recklessness is usually sufficient to establish liability for an intentional tort.[190] Returning to the factual scenario used throughout this book, an act will be reckless where B believes or suspects that consent has not been obtained (for the use of biological materials) from A, but proceeds to use the biological materials nevertheless. Intentional and reckless conduct covers an appropriate spectrum of behaviour without using the extreme mechanism of strict liability.

3 Elements

An individual's autonomy is the central interest that the common law ought to protect and liability should only extend to intentional and reckless conduct. The proposed tort may be characterised as the intentional or reckless non-consensual use of separated biological materials. This definition can be broken down into two elements.

First, biological materials must be separated from a living person. Any interference with materials that are not separated from a human body is protected by the tort of trespass to the person and battery. The second element of the tort stipulates that the defendant must use the biological materials with knowledge (or reckless indifference to the fact) that no consent has been obtained. This element raises several issues. What constitutes use? For example, is mere storage sufficient? Biological materials are likely to be used most frequently in research, as in the factual scenario analysed throughout Part II. B's use of X for research constitutes a non-consensual use of A's separated biological materials and would satisfy the second element of the tort proposed here. In the event that B has a reckless

[187] HTA 2004, s 5(1).
[188] See nn 93–111 above and accompanying text.
[189] *Three Rivers v Bank of England (No 3)* [2001] UKHL 16; [2003] 2 AC 1 (HL) 192 (Lord Steyn).
[190] Dugdale, above n 14, at 32–33.

disregard for whether A's consent has been obtained, then liability should also attach. For careless conduct, the law of negligence offers sufficient protection.[191]

A further issue concerns the level of information B must provide for A's consent to be legally effective. Under the proposed tort, the doctrine of consent would assume a primary role in determining liability, as it does in the context of battery.[192] In a criminal context, the HTA 2004 has recognised the pivotal role of consent in allowing the use of separated biological materials for the scheduled purposes covered by the HTA 2004. Drawing on the standards established under the HTA 2004, by analogy, A's consent would be sufficient if B disclosed to A that X was to be used for research purposes. Chapter four noted that the level of information required to be disclosed under the HTA 2004 has been determined by the Human Tissue Authority as set out in the Code of Practice on Consent.[193] The requisite standard of conduct required by the proposed tort could draw on the Human Tissue Authority's guidelines. In any event, the guiding principle should be that each individual has the right to control the purpose for which their separated biological materials are used. On the basis of this guiding principle, it is likely, for example, that B should be required to disclose whether X will be used for commercial purposes or for the creation of a cell-line. Assuming B discloses sufficient information to A so as to enable A's effective consent, then the second element of the proposed tort cannot be established.

The creation of liability also raises the question of remedies. In *Rees*, the House of Lords held that damages can be awarded for the infringement of an individual's autonomy. An analogy can be developed with *Rees*. As in *Rees*, an award of damages for the proposed tort should flow from the infringement of autonomy. The quantum of damages is a difficult question. It is unlikely, though, that a fixed sum (as was awarded in *Rees*) provides a satisfactory solution. There will be different factual scenarios. A fixed sum cannot hope to provide a level of compensation capable of general application because the quantum of damages should relate to the particular circumstances of the claimant. There is not an established body of jurisprudence in English law (or other jurisdictions) concerning the award of compensation for the loss of autonomy. Even cases that have awarded compensation for the tort of battery provide limited guidance.[194] In any event, in the factual scenario presented in this chapter it is likely that compensation would be relatively modest. As with the intentional infliction of emotional distress, an injunction (to prevent future misuse) would also provide further protection.

As acknowledged earlier, English courts are reluctant to establish new forms of liability. This hesitation is compounded by the fact that the UK Parliament had the opportunity to consider the creation of civil liability for misuse of separated biological materials (through the HTA 2004) but established only criminal sanctions.

[191] The concepts of negligence and recklessness do intersect and in some situations conduct may be categorised as both reckless and negligent.

[192] See nn 119–27 of ch 1 and accompanying text.

[193] See nn 29–34 of ch 4 and accompanying text.

[194] See, for example: *Mallette v Schullman* (1996) 67 DLR (4th) 321 (Ontario SC) [48].

D Conclusion

There is limited potential for English law to protect non-proprietary interests in separated biological materials. In relation to autonomy, article 8 of the EHCR and the law of negligence offer some protection. In relation to the protection of freedom from emotional distress, there is scope for the future development in English law of a separate tort for the intentional infliction of emotional distress. Here, American jurisprudence provides an example of how a tort of intentional infliction of emotional distress may offer some protection of non-proprietary interests, particularly in the context of interference with dead bodies.

In the alternative, this chapter considered the development of liability in English law to protect against the infringement of an individual's autonomy concerning separated biological materials. The principal difficulty here rests in securing recognition at common law that an individual retains a right to control the use and disposal of separated biological materials. Limitations in the English law of wrongs identified above reinforce the conclusion that property law provides the most appropriate framework.

8

Conclusion

Legal treatment of human biological materials in common law jurisdictions, including England, the United States and Australia, demands attention. The development of legal rules relating to separated biological materials has been haphazard and inconsistent. Legal uncertainty is inhibiting medical research and the ability of individuals to control biological materials separated from their bodies. To achieve the consistent development of legal principle, it is necessary to resolve the fundamental question: what property rights and non-proprietary interests are generated by separating biological materials from living persons and dead bodies?

Part I demonstrated that common law and statutory regimes in England, the United States and Australia fail to provide clear or coherent legal principles for determining the legal status of, and rights pertaining to, biological materials separated from dead bodies or living persons. Regrettably, this uncertainty informed the development of the Human Tissue Act 2004 UK, which, despite initial promises, does not establish a clear legal structure or regime of property rights.

Part II developed and articulated a rational foundation for the creation and allocation of property rights to separated biological materials. Presently, the work or skill exception operates as the principal legal mechanism through which property rights are created. Despite its prevalence, the exception should not be adopted as the governing principle for the recognition of property rights. Not only is the exception a misapplication of the specification doctrine, more importantly, it is incapable of general application. In contrast, the 'detachment principle' provides a logical explanation to support the creation of property rights. In the case of biological materials separated from a living person, property rights should be allocated to the person from whom those materials are removed. In the case of a dead body, those rights should be allocated to the deceased's representatives. The 'detachment principle' also offers a foundation for the application of property rights to a variety of additional biological materials created and used during scientific and medical research. These materials include cell-lines, which are increasingly utilised by institutions such as the UK Biobank and UK Stem Cell Bank.

Significantly, the normal incidents of ownership include the right to sell the property. A tide of legislative policy is running against individuals selling biological materials. These policy considerations are important and require examination. Further empirical studies are necessary, for example, to assess the impact that an individual's right to sell their biological materials is likely to have on scientific and medical research.

Nevertheless, such policy considerations are secondary issues that concern the content of property rights and not the creation and allocation of property rights which have been the focus of this book. The law must initially determine who has the legal right to control separated biological materials. Only after an adequate answer is provided to that question can such policy considerations be identified and addressed.

This book articulates and supports two important conclusions. First, the law should be developed so that individuals (or their representatives) have the right to control biological materials separated from their bodies. Secondly, such control should be granted through the legal recognition of property rights. This approach provides a structure of rights which is necessary to address the difficult questions that arise when established legal principle confronts ongoing scientific development. Only then can parliamentarians, courts, medical researchers and the wider community successfully navigate the intersection between law and the human body.

INDEX